Successful Project Management

Successful Project Management

A Step-by-Step Approach with Practical Examples

Fourth Edition

Milton D. Rosenau, Jr.
Gregory D. Githens

WILEY

John Wiley & Sons, Inc.

For general information on our other products and services please contact our Customer Care Department within the United States at (800) 762-2974, outside the United States at (317) 572-3993 or fax (317) 572-4002.

Wiley also publishes its books in a variety of electronic formats. Some content that appears in print may not be available in electronic books. For more information about Wiley products, visit our web site at *www.Wiley.com*.

Library of Congress Cataloging-in-Publication Data:

Rosenau, Milton D., 1931–
    Successful project management : a step-by-step approach with
practical examples / Milton D. Rosenau, Gregory D. Githens.—4th ed.
        p. cm.
    Includes bibliographical references and index.
    ISBN-13: 978-0-471-68032-1 (cloth)
    ISBN-10: 0-471-68032-X (cloth)
    1. Project management.   I. Githens, Gregory D.   II. Title.
    HD69.P75R67 2005
    658.4'04—cd22

                                        2005005180

Printed in the United States of America

10 9 8 7 6 5 4 3 2 1

Contents

Preface

WHO THIS BOOK IS FOR

This book is for anyone interested in a pragmatic approach to managing projects and programs. We have found that the material is valuable as a refresher for the experienced manager and as a primer for the person who wants an introduction. The factors that lead to project success are known and knowable regardless of the industry, the size of the project, or its technology. Good project performance can be ensured with the skillful application of the processes, tools, techniques, and concepts of project management.

A natural and primary audience for this book is the person named as "project manager." (In some cases, the label may be "project leader," "project engineer," or similar variants.) It is common to find individuals who have been trained in a technical skill (e.g., engineering, science, accounting, and programming) thrust into management roles with little training, coaching, or mentoring. We often will address the reader as "you" in recognition of this important audience.

Another important audience is project team members. Over the years, we have seen many situations where the project manager was well trained in the tools and principles of project management but became frustrated when he or she tried to engage the project team to help them apply the tools. Projects are a collaborative activity. This book will help team members understand their roles and responsibilities in supporting the development and execution of project plans.

In recent years, the "process view" of the enterprise has shown the value of improving performance of complex work activities. It is the job of senior management to help create the system that allows for the consistent and systematic development and delivery of projects. We wrote this book so that people with only limited time but with a need for a strategic perspective can identify the success factors for delivering projects. Thus, another important audience for this book is executive sponsors and customers as well as functional managers.

Many organizations now have adopted the concept of a project office. This book will provide important insights to those people who are organizing or operating a project management office.

THIS BOOK'S APPROACH TO PROJECT MANAGEMENT

This book is useful for any type of project, regardless of size, technology, or industry. In addition, we address portfolio management and program management of integrated collections of projects.

We organized this book to provide a simple model of the fundamentals of project management. Our straightforward approach is based on a combined 70 years of experience with new product development for consumer and industrial markets, chemical formulation, engineering, government contracts, research, management consulting, and volunteer organization projects.

This book divides the management of projects into five general managerial functions and emphasizes the importance of integration, as illustrated in Figure P-1.

1. *Defining.* Defining the project's goals.
2. *Planning.* Planning how you and your team will satisfy the Triple Constraint (goal) of performance specification, time schedule, and money budget. The plan depends on the mix of human and physical resources to be used.

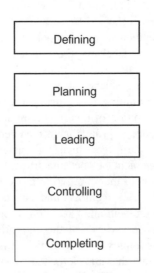

> Defining

> Planning

> Leading

> Controlling

> Completing

FIGURE P-1. *The five activities are different but interdependent.*

3. *Leading.* Providing managerial guidance to human resources, subordinates, and others (including subcontractors) that will result in their doing effective, timely work.

4. *Controlling.* Measuring the project work to find out how progress differs from plan in time to initiate corrective action. This often leads to replanning, which may force a goal (definition) change, with a consequent need to change resources.

5. *Completing.* Making sure that the job that is finally done conforms to the current definition of what was to be done, and wrapping up all the loose ends, such as documentation.

Although these are distinct, they are interrelated, as shown by the arrows in Figures P-2, P-3, and P-4.

The first two steps are not necessarily separate and sequential, except when the project initiator issues a firm, complete, and unambiguous statement of the desired project output, in which case the organization that will carry out the project may start to plan how to achieve it. It is more common to start with a proposed work definition, which is then jointly renegotiated after preliminary planning elucidates some consequences of the initially proposed work definition. The definition must be measurable (specific, tangible, and verifiable) and attainable (in the opinion of the people who will do the work) if you want to be successful. Being successful also requires that management agrees that the project is justified and that the resources the project team needs will be available.

Thus, in fact, the resources to be dealt with in the leading phase often must be considered before planning can be finished (Figure P-2). For instance, you might need engineers familiar with carbon fibers if the plan for a materials study project includes the study of that kind of material, whereas you would use a metallurgist if the project were to study only metals.

No project goes in accordance with your plan. What you don't know when you start is where it will go awry. Consequently, as you will see in later chapters, replanning is almost always required, thus frequently amending the negotiated def-

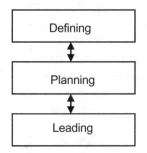

FIGURE P-2. *Defining, planning, and leading activities often must be considered simultaneously.*

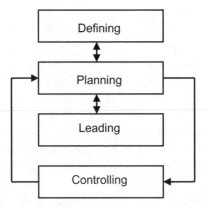

FIGURE P-3. *Controlling (or monitoring) is carried out to detect deviations from planning.*

inition (Figure P-3). Ultimately, the project can be completed when the work that is done satisfies the current requirement (Figure P-4).

Nevertheless, the five-step managerial activity process covers each required action and is a useful conceptual sequence in which to consider all project management. Thus, this book is organized according to it.

Each chapter is short and can be absorbed in 1 to 2 hours. The chapter sequence is a good match for the chronological concerns during a typical project. We caution you, however, that there is no single "cookbook" or template to follow. We encourage you to scan the chapter highlights and use the index.

Because projects are complex and strategic, they require an appropriately sophisticated set of managerial tools. A tool icon is inserted at points where

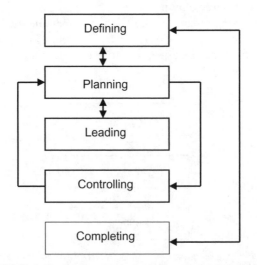

FIGURE P-4. *Completing depends on the current defining basis.*

we describe what we consider to be a particularly useful tool that you may want to use. Some of these are simple to use, and some will require practice. Perhaps a few will never be right for your management style. Because not all these tools will be useful for you in specific situations that you encounter, you will have to pick and choose which tools to use when.

We tried to use graphics and examples liberally. This book contains several illustrations of computer project management and other software outputs, most from the widely used shrink-wrap project software package Microsoft Project. We are not endorsing this product, nor are we discouraging its use. There are other widely used and effective single user and enterprise computing systems. Our goal is to explain a few of the key useful aspects of this class of software. Employing such software will not make a person a successful project manager, and using it is not the same as being a project manager. Software is a tool to help but not a solution, especially to "people" problems. Nevertheless, you can be a more effective project manager if you employ such software in situations where it will be helpful to you.

CHANGES IN THIS FOURTH EDITION

The first three editions of *Successful Project Management* proved the value of the book's approach to helping readers improve project management skills. In the two decades that followed the publication of the book's first edition, people have increasingly come to regard project management as a "profession" instead of a "job." A short list of developments in the field of project management would include the process view, virtual teams, new theories and practices of motivation, the quality view, project management offices, and so on. Practitioners have developed and documented a recognized body of knowledge and an ability to certify individuals and to recognize organizational capability (also called *maturity*) in the process of project management.

Reflecting the explosion of documented project management knowledge and standards, we have made extensive changes to this fourth edition. This edition clarifies some of the previous material, brings it up to date, and eliminates some material made obsolete or irrelevant by the growing sophistication and professionalization of the project management field.

We rewrote a number of chapters to bring these up to date with contemporary concepts, standards, and practices of project management. We have incorporated and reflected the standards that groups such as the Project Management Institute (PMI) have publicized. Throughout this book, we have used standard language, enhanced and corrected the graphics, and recognized the use of enterprise computing, networks, and the World Wide Web.

Readers familiar with the earlier editions of this book will notice that there have been substantial changes to the book. Five of the first six chapters have been rewritten to better describe the role of integration in project planning. Chapter 11 on risk and issues management is also completely rewritten. Moreover, all the remaining chapters have been augmented to reflect contemporary thinking and practices in the project management field.

Greg Githens joins Mickey Rosenau as coauthor for this fourth edition of *Successful Project Management*. He brings a substantial background as a project management practitioner, consultant, trainer, and professional contributor to the project management field.

PROJECT MANAGEMENT AS A DISCIPLINE

It is an exciting time to be in project management. There has been an explosion of knowledge and tools for the field and increasing recognition of "good" and "bad" practices. Organizations who have embraced it are achieving outstanding results. Yet, successful project management has been and will be based on people. Project management is a *discipline,* a word that has its semantic roots in the ideas of teaching and learning. As an individual and organizational competency, project management discipline involves leadership from individuals who have the personal backbone to withstand the criticism of undisciplined, impatient people. It requires an organizational commitment to investing sufficient up-front time and to involving other people, recognizing that different points of view result in more creative, optimal outcomes.

Milton D. Rosenau, Jr., CMC, FIMC

Certified Management Consultant
Rosenau Consulting Company
Bellaire (Houston), Texas
mrosenau@houston.rr.com

Gregory D. Githens, PMP, NPDP

Managing Partner
Catalyst Management Consulting, LLC
Findlay, Ohio
GDG@CatalystPM.com

About the Authors

Milton D. ("Mickey") Rosenau, Jr., CMC, FIMC, heads Rosenau Consulting Company, which he founded in 1978 following a 21-year career with industrial and consumer products companies. He is the author of dozens of publications and nine books, including *Successful Project Management* (3d edition, Wiley, 1998) and *Successful Product Development: Speeding from Opportunity to Profit* (Wiley, 1999). He was a past president of the Product Development and Management Association (PDMA) and was editor-in-chief of the *PDMA Handbook of New Product Development* (Wiley, 1996).

Gregory D. Githens, PMP, NPDP, is managing partner with Catalyst Management Consulting, LLC, a management consulting firm specializing in project management and new product development. His clients have achieved improved time to market, better metrics, better strategic alignment, and improved risk management, among other benefits. Mr. Githens has been a frequent contributor to the profession, including developing professional standards, writing over 30 articles, and public speaking.

1

Projects, Project Management, and Program Management

Projects are a kind of work that is temporary, unique, and progressively elaborated. Accordingly, project management is a discipline that includes a specific body of knowledge as well as a specialized set of tools. In this chapter, we explain how project management is different from process management and ad hoc management, the nine knowledge areas—stressing the importance of integration and managing expectations— and overview five managerial functions.

Project success doesn't just "happen"; it comes from people using commonsense tools that are suited for the special nature of projects and applied in an organizational environment that accepts discipline and rigor. To understand what makes project management "successful," we need to start with its basic unit, which is the project. In this chapter, we will explain what a project is and isn't and describe the foundations of project management as a discipline.

PROJECTS ARE A TYPE OF WORK

It is important to understand what a project is so that the project manager and project team can select appropriate project management tools. This section provides a basic definition of a project.

First, let's examine some characteristics of *any kind* of work activity, including projects. Thus, *all work,* including projects:

- *Uses resources.* For the purpose of this definition, resources include people, capital, equipment, ideas, and so forth. Whether the organization is refining

1

oil, building a building, programming a computer, conducting a management consulting assignment, designing an instrument for a satellite, developing a new product or service, or surgically removing a cancerous tumor, a manager is responsible for the effective application of resources.

- *Is requested or needed.* Customers, and their willingness to spend their scarce money for goods and services, are the lifeblood of any organization, be it government, business, or charity. Successful organizations pay attention to customer needs to deliver goods and services that customers' value.
- *Has goals.* Generically, management is a process of establishing goals and directing resources to meet those goals.

These three factors are descriptive of projects but are not sufficient to distinguish a project from a nonproject. The accepted definition of a project is that it is a temporary work effort that produces a unique result. Let's look at each of the three characteristics that distinguish projects from other kinds of work:

- *Projects are temporary.* Temporary means that there is a beginning and end to the project. Projects start when the sponsoring organization authorizes the project, and projects end when the project meets the requirements. All well-managed projects must come to an end! For example, the project of constructing a major downtown hotel would take one or two years to complete, but the project would complete the work.
- *Projects are unique.* Unique means that the work product or processes that create it are novel or different. Even though a second software project to write an accounts payable system is very similar to the first such project, there will be some differences, perhaps something as simple as the format of reports. The same is true of digging two ditches (the purpose or terrain may vary) or organizing two conventions (the sites or programs may differ), and so on. For example, while hotels may have similar layouts ("footprints"), the people and materials involved in the construction are different for each hotel.

A project is a temporary work effort that produces a unique result.

- *Projects are progressively elaborated.* This means that a project proceeds in steps or stages. Most well-managed projects use a phased approach, where the project defines the phases according to its control needs. For example, real estate developers often acquire land speculatively and then put together deals to construct hotels, restaurants, and convention centers according to the needs of the local market. We will describe more on the project life cycle phase in Chapter 5 and later chapters.

Now let us see how a typical organization might use this definition of projects as temporary, unique, and progressively elaborated to identify work activities that would best benefit from the project management tool set. Figure 1-1 shows three

	Definitely *Is* a Project	Might Be a Project	Definitely *Is* Not a Project
Example	Constructing a hotel.	Painting a bedroom	Processing employee time data to produce payroll checks
Rationale	Meets the generally accepted definition of a project. It is a one-of-a-kind product It started from a piece of ground and was developed into a complex.	Is probably better to call this work a task because of its simplicity. Extensive formal documentation, status meetings, and so forth are probably not necessary.	Better thought of as process or operations management. Making a major change to upgrade the payroll system *could* become a project.

FIGURE 1-1. *Identifying a project, nonproject, and possible project.*

types of work that might take place in an organization. The first and third columns are straightforward. To define what something *is*, it is helpful to define what it *isn't.* Let us look at the column headed, "Might Be a Project." The "Might Be" column is important in addressing a common complaint about project management, which is project management is bureaucratic, involves many meetings, and so forth. The difference (and need for sophisticated project management) arises from the need to manage across interfaces and deal with complexity.

Why should you care about distinguishing projects versus nonprojects? Not everything individuals or organizations do is a project, but *some* things are a project. The things that are projects are typically not the day-to-day work of people but have to do with creating a new outcome. Projects are strategic! Because projects are complex and strategic, they require a particular and sophisticated set of managerial tools.

Not everything individuals or organizations do is a project, but some things are a project

While the term *project* refers to work activities, people also commonly use the term to refer to organizations of resources. Hence, projects perform work: After initiating the project, they "plan their work and work their plan," make changes as necessary, and close out the project.

Contemporary thinking identifies projects as essential components of enterprise strategy. Projects are one important kind of organizational work because they create change. Because projects create change, good organizations explicitly align their projects with the investment policies and intention of management. We explore the selection and definition of projects further in the next and subsequent chapter.

PROJECTS DISTINGUISHED FROM TASKS AND FROM PROCESSES

Projects are different for other work activities and require different tools.

It is important to distinguish projects from other kinds of work. A project is *more complex* than a task and *more unique* than a process. There are not clear-cut distinctions, and each individual and organization will develop and apply judgment on where and how to clarify the differences.

Projects are more complex than tasks. For an individual, going to work on some mornings might seem to be a major undertaking. It is something that the individual must do in order to maintain employment, it requires resources, and it has a goal. The individual needs to apply some forethought to identify the best route considering the risks involved. He or she could even create a "to do" checklist for the undertaking to help him or her remember all the steps.

However, from the perspective of an enterprise, an individual's commute to work is a task and not a project. It does not involve the coordination of people (although organizing a car pool could be an exception), does not require capital investment, and does not benefit from managerial oversight. In most organizations, a task is something that an individual can accomplish by himself or herself in a few hours or a few days of time.

While both projects and tasks have an end-state goal and use resources, there is little value in developing a project management approach for the *task* of going to work. We think it is important to include in the definition of a project—at least for purposes of this book—those endeavors where project management tools add value. There is no sense in adding the sophistication of project management tools for work that simply does not require sophistication.

We want to stress that project management is *not* managing your "to do" list. In Chapter 6, we again turn to the discussion of tasks but examine them as the activities that the project organizes into a work breakdown structure.

Considerable progress has been made in identifying the factors that lead to successful project management.

Some organizations fail to recognize the distinction between tasks and projects. An *ad hoc* approach is suitable for a task but not for a project. In these organizations, a mixture of effort and luck drives performance. Organizations get inconsistent results from their projects and tend to attribute the result (good or bad) to the individual. In the past 10 years, however, there has been a growing movement within the project management profession to measure and develop "project management maturity" systematically at the enterprise level. It is now much easier to identify the causes and consequences of successful project management.

Now let's examine how projects differ from processes. Processes have three components: inputs, transformations, and outputs. From an organizational standpoint, processes are mostly repetitive and produce common outputs. Projects are different from processes because they have less consistency in inputs, transformations, and outputs. For example, if the individual's project is to build an amplifier circuit, at some point, building a second, third, or fourth amplifier circuit ceases to be a project and becomes a repetitive activity. If each amplifier is virtually identical, we have a production line; thus we are managing a process, not a project. The lesson here is that the individual should determine if the requestor's requirement is to build a single amplifier, or to build a batch of amplifiers, to build an amplifier production line. Often, the individual performs unwanted or unbudgeted work, a phenomenon known in the project management community as *scope creep*.

Refer back to the right-hand column of Figure 1-1. Other examples of processes are manufacturing, payroll processing, and building maintenance. Process management focuses on standardization, particularly of the output. To achieve consistent, high-quality, standardized outputs, process management places requirements on the inputs (the raw materials) and the production that transforms the inputs to the outputs. High-volume, high-quality output is typically a goal of process management.

The disciplines of "process management" and "project management" differ in goals and metrics. In process management, goals and metrics are set up to eliminate variation within the process because variation is wasteful. A new product development example is instructive on this point. If a person is purchasing an automobile, he or she assumes that any car that he or she purchases will be consistent with other like models and will meet the advertised performance specifications. Managers design the manufacturing process to eliminate variation in order to produce automobiles of an expected, consistent quality. Henry Ford's famous quotation about the Model T, that "a customer can have any color that they want as long as it is black," is an extreme example of the efficiency mind-set. On the other hand, customers desire variety and have requirements for an automobile with features and functionality that make it distinctive, for example, new and different colors. Ford's competitors were able to create distinctiveness that the customer valued and gained market share in part because of Ford's rigidity in the use of process management and metrics.

Thus, projects allow organizations to give customers new and value-added choices. In this sense, variation is good, because customer-perceived value is a source of competitive advantage.

In recent years, the project management literature has contained considerable discussion of the process view of project management. Projects *do* involve repeatable activities such as capturing requirements, building teams, and publishing reports, but the processes are not high-volume, long-term production lines. Except for very large aerospace and construction projects, projects seldom perform high-volume work activities. More typically, projects use process management to develop routines so that people can manage frustrations or focus on creative tasks. For example, project status meetings are a repetitive activity within projects. Project status meetings can benefit from standardizing on agendas, goals, time, and so forth.

Then the project effort can focus on creating unique and sometimes first-of-a-kind items.

Finally, don't confuse a procedure with a process. A *procedure* is a job instruction for accomplishing an operation. For example, in a chemical processing plant, there are standard "lock-out, tag-out" rules for shutting off equipment or entering a confined space. Technical and process-oriented organizations have volumes and volumes of procedures for people to follow. Some people often use the term *methodology* to describe a system of procedures. These procedures are necessary because of the following:

- There is a considerable amount of detail that a person must remember.
- The operation must be performed in a specific sequence.
- Failing to complete *all* steps in the proper sequence could cost lives or significant money. In some cases procedures are subject to government regulation and oversight, for instance, in the development and manufacture of a new drug or medical device.

It follows naturally that organizations would want to exert some similar types of control over the initiation, planning, execution, controlling, and closing processes of their projects.

Attempts to control projects through procedural control seldom work as well as the designers of the control process desire.

However, these attempts to control projects through procedural control seldom work as well as the designers of the control process desire. Once the documents are written, organizations place them in a library to indoctrinate people in the procedures. However, in practice, people do not pay attention to the procedures and soon start ignoring them. When people deviate from the rules (often for a very good reason), the "methodologist" typically writes a new rule. This expansion of "methodology" can grow to a multi-volume set that people view with cynicism. For example, one organization developed a binder of procedures for new product development activities that was so thick and heavy that people developed the slang name "Thud Document" because of the "thud" sound the book made when a person dropped it on a table.

Here are a few commonsense observations about the difference between a process and a procedure:

- Projects have many unique facets, so many of the procedures do not apply or only apply partially.
- Projects are more complex than tasks, so projects require knowledge of many things.
- People have limited capacity to absorb abstract information.

- People are under pressure to get to work and deliver visible results quickly.

The best organizations avoid a rigid set of step-by-step procedures for project management. Instead, the best organizations educate all stakeholders on the principles and allow for discretion and common sense. To be sure, templates and checklists are helpful job aids for the novice; just don't become a slave to your tools.

> *Project management is fundamentally a commonsense approach that depends on the good judgment of people.*

PROGRAMS ARE COLLECTIONS OF PROJECTS

A program is a collection of projects grouped together to get advantage from their combined management. For example, the trans-Alaska pipeline and the manned lunar landing projects required many years and billions of dollars. The overall success of such programs depends on hundreds or thousands of projects. Programs usually are larger than projects and are collections of projects that come to an end when the sponsoring organization makes a determination to end the program *and* when the projects that make up the program are completed.

> *A program is a collection of projects grouped together to get advantage from their combined management.*

Better organizations make a distinction between *program* and *project* and between *program management* and *project management.* These are not interchangeable terms.

The reader should have some awareness of a phrase related to program management that is a set of practices for understanding the relationship of projects to an organization's strategies: *project portfolio management.* People and organizations use this phrase to portray projects as assets invested by the organization to achieve its goals. Project portfolio management includes concepts such as the following:

- Setting priorities across all projects
- Allocating resources across all projects, including project managers and project participants
- Project selection and deletion
- Understanding that projects are different and require different approaches (For example, there are capital equipment projects, innovation projects, regulatory compliance projects, system support projects, and so forth.)

If you work for a large organization, you should find out if your organization has a project or project management office and consult with its leaders to understand how your own organization defines projects, programs, and portfolios. In addition, find out what kinds of job aids, training, coaching, and support are available to project managers and project participants.

PROJECT MANAGEMENT MATURITY

Humankind has performed and managed projects since the earliest of times. People have looked on the artifacts of projects—both from earlier civilizations and from contemporary periods—with amazement. Was "project management" used to create these wonders? In a sense, yes. History shows that people used many advanced insights in their planning and logistical practices. Still, we can't help but visualize a brutal taskmaster with a whip standing over cowering laborers and know that this type of coercive power would be an inappropriate "people skill" in today's organizations.

The project manager is not the schedule mechanic.

In the three decades that followed World War II, organizations found significant economies of scale and centralized massive "mainframe" computation capabilities, leading to the development of sophisticated techniques for task scheduling, estimating, and resource management. The "project manager" often was the only person with the training and desire to use this system, which unfortunately caused many people to regard the project manager as the "schedule mechanic." The use of information technology continues to be important to project management because it has changed and even eliminated many of the middle management roles. Organizations have decentralized steadily and have become more networked. Now, any person with access rights can view and update the project documentation instead of waiting for a directive from their manager. Information technologies have allowed all project team members to multitask and contribute to a number of work activities in a number of different ways. This decentralization has led to more empowerment and more accountability for results at the individual and team level of the project. The project manager is the person we look to help the team achieve the results.

In recent years, it has been popular to say that project management is the "accidental profession," but that is changing. Most project managers did not start their careers as project managers. Most people get exposure to projects early in their careers as technical contributors, and gravitate (or are pushed) into taking on a more systematic view of projects. As they gain experience with projects, individuals often decide that the project management career path is more to their liking than the technical path.

Now, in the twenty-first century, there is a solid and global recognition of the importance of project management as an explicit organizational activity with best practices and skills. One source of support for this statement is standards developed and publicized by such groups as the Project Management Institute (PMI). Through PMI and other similar organizations, people increasingly have developed and documented a recognized body of knowledge and an ability to certify individuals and to recognize organizational capability (also called *maturity*) in the process of project management. At the university level, there is a growing understanding and dissemination of the principles and practices that make for successful project management. *Project management* is the application of knowledge, tools, and techniques to an individual project. People also commonly use the term to respond to the enterprise's capacity and capability to deliver projects.

What does this mean to the various project stakeholders? First, individuals should give serious thought to expanding their knowledge of the profession beyond what we describe in this text. Second, individuals should consider including a stint in project management in their career path, especially if they aspire to top management positions in their organizations. Third, project management is not simply a cadre of competent project managers. Successful project management is also a result of a systematic organization-wide capability. Many leading organizations have and are investing significant resources to ensure that they manage projects well.

Thus, the contemporary view of project management is that it is larger than a "tool" that individuals use. Project management is a discipline: There is an active and growing cadre of people researching and documenting the practices and issues associated with the success of projects.

INTEGRATED PROJECT MANAGEMENT

This section describes a contemporary and systematic view of project management. There are many published global bodies of knowledge. This book will adopt the model published by the Project Management Institute (PMI) in its reference, *A Guide to the Project Management Body of Knowledge.*

The project management profession recognizes nine areas of project management knowledge. Review Figure 1-2, which shows the nine elements of project management, noting that the center of the diagram is integration, which will be our starting point. One of the key points to remember in this book is this: Project management is *primarily an integrative activity.* Here are some reasons for making this claim:

> **Integration is a process of synthesizing different concepts into a unified whole.**

- Most product failure is at the interface of subsystems, and individual contributors typically do not pay attention to interfaces until after they have completed most of their work on their assigned piece of the system.
- Time and again, the biggest complaint of people in projects is poor communications.
- Senior managers expect project managers to manage their projects as if they were a business, taking a strategic and holistic perspective on the project.

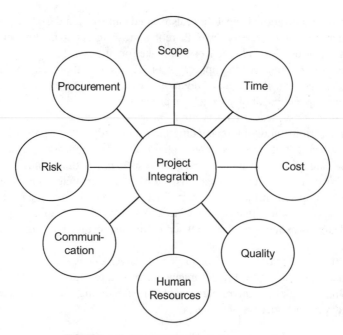

FIGURE 1-2. Integrated project management.

Within a project, there are three integrative processes (listed below). When managing a single project, the team performs integration with respect to the other knowledge areas. This includes the following:

- *Developing the project plan.* Integrating the requirements that comprise the product scope with the activities that comprise the work scope while establishing plans for cost, time, quality, risk, communications, human resources, and procurement.
- *Executing the project plan.* Now that the project has "planned its work," it is now time to "work its plan," and the execution must be systematic.
- *Managing the changes to the plan.* All projects encounter changes that are different from the assumptions on which the plan was based. This change needs to be managed, and in an integrated way.

Of course, there are integration needs: the product into the established systems and the project with other work in the organization. We will touch on this type of integration thoughout this book.

We hope that we have convinced you that project integration is significant. Recall that many people enter project management "accidentally." Experience shows that most of these new entrants tend to stay in their comfort zone. These three assertions are fundamental to *Successful Project Management:*

- Integration is at the heart of project management and distinguishes the excellent project manager from the mediocre one.
- Integration expertise is as important as technical expertise.
- Integration adds value to the customer.

The discussion of the nine knowledge areas is a useful foundation for understanding the vital concept of successful project management: balancing competing demands. We will describe the concept of balancing competing demands in Chapter 2 and elsewhere in this book. The paragraphs that follow describe each of the remaining eight knowledge areas in more detail.

Project Scope Management

In project management, project managers use the word *scope* to indicate that there is a boundary where things are "in scope" and "out of scope." There are three kinds of scope, as we describe in the following paragraphs.

The *problem scope* is the definition of the problem or opportunity, often called *requirements*. Many frustrations that project managers and participants face originate with poor requirements, suggesting that many organizations need to improve the ways they set boundaries over what is "in" and "out" of the problem scope. We will take up this challenge in Chapter 3.

The *product scope* includes the features and functionality of the product of the project. The product scope is the "solution" that satisfies the customer's needs, wants, and requirements. In Chapter 3, we will discuss how product requirements set the boundaries for the design of the product scope.

Product is synonymous with *result,* so don't assume that the term *product* means a tangible item. The term *product* means any result—tangible or intangible—delivered by the project to the user. Here are three examples: The result of a research and development (R&D) project into the cellular behavior of Alzheimer's disease is new scientific knowledge, the result of a project to change hospital visiting hours is better patient and visitor satisfaction, and the result of a project to educate people on cancer prevention is the increased appearance of healthy behaviors.

Project scope and *work scope* are terms describing the work that creates the product that satisfies the requirements. As we will explore further later, the "project is the work that creates the product of the project." Project managers manage work. Completion of the product scope is measured against the requirements, whereas completion of the project scope is measured against the plan. In Chapter 6, we will describe how the work breakdown structure (WBS) is the measurement tool for project scope.

Project scope management is concerned primarily with defining and controlling the work that "is" or "is not" performed. Because scope is an ambiguous concept, we recommend using a modifier in front of the word *scope:* Be clear whether you are talking about *problem* scope, *product* scope, or *work* scope.

Project Time Management

Time is a very visible project dimension. Time is easy to understand (as evidence of its ease, people learn how to read a calendar at an early age), customers are demanding, and people are impatient. Hence, projects must make skilled use of the tools and techniques that generate a project time schedule. The techniques and skills of project time management include defining the work activities, sequencing the activities, estimating the duration of the activities, and integrating these into a time schedule. During execution, the project needs to manage work to the schedule and make appropriate changes.

In the ideal world, the customer and sponsor define the problem scope and the product scope before asking for a schedule. Unfortunately, the more common experience is that dates for project completion are set before the individuals participating in the project understand the problem and product scope. We elaborate on this in Chapter 25, especially in Figures 25-6 through 25-9. Sometimes people view the job of the project manager as being the "schedule mechanic," but this is a too-narrow role that does not ensure that projects are successful. Good project managers manage time expectations well but in balance with other performance parameters.

Project Cost Management

Projects typically measure cost in terms of some unit of currency, but it is important to remember that the currency is just a metric for some kind of underlying resource. For example, an individual's time is a resource, and it has value. Project cost management thus has to do with estimating, valuing, and managing resources.

In this knowledge area, projects need to develop and manage two related processes: cost estimating and budgeting. Cost estimating includes planning the resources needed to accomplish the project work scope, pricing the resources, and generating a total project price (or cost). The project budgeting process is different from the estimate. The budget is the baseline plan for the cost, by which the project will measure its progress and its variances. Of course, as the project work scope changes, the budget may change, and the project manager needs to manage this change.

Project Quality Management

Like the definition of project scope, the definition of project quality is ambiguous. Consider that there are at least four ways that people judge quality, and people feel passionately that they have the "right" definition of quality:

- Satisfaction of product requirements
- Conformance to standards and policy
- Absence of product defects
- Delighting the customer

Thus, one of the important activities that take place in a project is defining quality for the product and for the project. Generally, it is best to define quality in terms of the customers' needs (both expressed and unexpressed). The project manager leads the project quality effort through quality planning, quality control, quality assurance, and quality improvement. Some projects have elaborate quality systems in place to ensure quality, whereas others are less formalized.

Project Human Resources Management

Projects involve people, and the techniques and skill applied to leading and managing people significantly affect the project's performance. Project human resource management involves a number of activities, including organizational planning, staff acquisition, and team development.

Good project managers make effective use of the people involved with the project and treat them as important stakeholders. One measure of project success is that participants feel like their involvement was a worthwhile experience.

Project Communication Management

The intent is the proper generation, collection, dissemination, storage, and ultimate disposition of information. It provides the critical linking of people, ideas, and information that is necessary for success. Different project stakeholders have different communication needs.

The component processes of project communications management are communications planning, information distribution, progress reporting, and administrative closure.

Project Risk Management

As we wrote in the first line of this chapter, successful projects don't just happen. Regrettably, poor project performance occurs frequently. Risk management is at the heart of successful project management and deserves considerable attention. Project risk management includes identifying, analyzing, and responding to project risk. Risk-response planning includes defensive actions such as risk avoidance, risk mitigation, risk transfer, and risk acceptance (which includes the passive response of "dealing with it when it happens" and active acceptance—the use of contingency planning).

Often practitioners include issues management as an associate practice of risk management. We will cover both risk and issues management in Chapter 11.

Project Procurement Management

One of the most pronounced current trends in project management is the use of contractors, vendors, and partners to accomplish significant pieces of a project's work. Project procurement management is concerned with the processes for acquiring goods and services from outside the performing organization. A good project needs to pay attention to procurement planning, solicitation planning, source selection, contract administration, and contract closeout.

THE PROJECT MANAGEMENT "HAT" IS DIFFERENT FROM THE TECHNICAL OR PRODUCT MANAGEMENT "HAT"

In the preceding paragraphs, we described the importance of project integration and the knowledge areas that comprise project management. We hope you are recognizing that the knowledge and skills of project management cover a broad area and that there is much to know. Start with this simple principle: The roles needed for project management work are different from the roles needed for technical work.

Most experienced project people agree with this statement: "The best technical people seldom make the best project managers." Why? For some, managing people is the most difficult aspect of managing a project, especially for recently appointed managers whose academic training is primarily in a technical discipline such as engineering, computer science, or even construction management. Such people tend to be more comfortable with things and numbers than with people.

Project management is not technical work. There is a significant difference between product-oriented work and project-oriented work. Product-oriented work involves applying technical or "domain-specific" knowledge to creating a product. For example, the knowledge necessary to create computer code, design a circuit board, dig a foundation, or create a graphic is *product-oriented* knowledge.

The project is "hat" is different from the technical "hat."

Project-oriented knowledge involves the nine previously described knowledge areas. Project management is an integrative activity. Good project managers avoid the expert's propensity to concentrate on the technical or quantitative aspects (e.g., engineering analyses or task budgets)—and instead become more oriented toward making things happen through people.

* Most project managers must provide both technical expertise and integration expertise. They need to perform the project in order to develop the product! Here is a tip that has helped many technically trained project managers. At any given time, stop and ask yourself: What hat am I now wearing? The project manager hat? The technical hat? Some other hat? People find that if they can identify the role they are playing *at any particular time,* they can select the appropriate behaviors for that role.

EFFECTIVE PROJECT MANAGERS MANAGE EXPECTATIONS OF STAKEHOLDERS

Projects involve multiple organizations, for example, different departments on a cross-functional new product development team, different businesses on a shopping

---

*The tool icon is used next to a paragraph that describes a particularly useful tool that you may want to use. See the preface, page xvi, for more information.

mall joint venture, or even public-private partnerships for a charity fund drive. Each of these organizations has multiple goals, functions, and missions at any given moment if for no other reason than that each is composed of many individuals with varied skills, interests, personalities, and values.

A *stakeholder* is a person or organization that has an interest in a project or the outcome of a project. Figure 1-3 illustrates the project manager at the center of a web of stakeholders. Each stakeholder has differing values, missions, and aspirations that become the foundation for different expectations about the definition of project success. Experience shows that each stakeholder will evaluate the project's performance independently. For example, a company decides to fully absorb an operating division that previously was independent in order to gain efficiencies of scale in operations. Economic, political, and social

> **Project management success means actively managing the expectations of various stakeholders.**

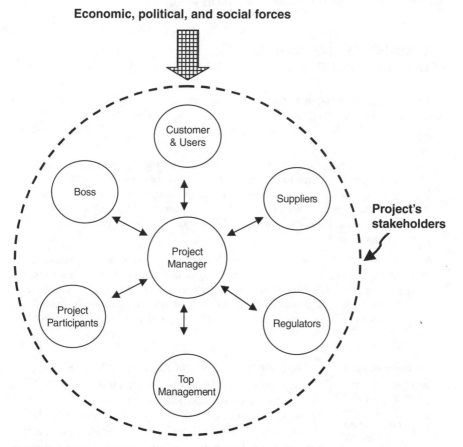

FIGURE 1-3. *The project manager must manage expectations of stakeholders, where each has differing values, and missions reflective how they perceive the economic, political, and social environment.*

forces may be external to the company, too, such as industry restructuring to "outsource" organizational functions (e.g., operating a help desk) to a specialized firm. Individuals (you as project manager, your boss, others) are affected by changes in tax rates and laws that may be altered during a project. If your country (or the country in which a key project participant is located) declares war on another nation, that potentially will affect all parties. Thus, project success is subjective (in the eye of the beholder), and an important task of the project manager is to manage stakeholder's expectations, a point we will emphasize throughout this book.

All project managers have experienced the frustration of dealing with the many other directions in which the organization seems to be (and often is) moving. People act logically based on the way they see the data. Because each stakeholder perceives the situation differently, each develops a set of beliefs and behaviors that will cause differences in opinions and style. Project management, in large part, is the management of interpersonal conflict, which is inherent in complex organizational situations. Part 3 of this book will cover ways to deal with this "soft stuff," which one sage observer noted was often the "really hard stuff."

A ROADMAP OF FIVE IMPORTANT PROGRAM MANAGEMENT FUNCTIONS

Project management requires many different managerial activities. It is helpful to have a way to orient oneself.

Consider the predicament of one project manager.

I am a project manager working with four or five other people. Our project consists of redesigning a piece of hardware, making any improvement possible, and then making five to ten prototype systems. Our customer has not specified any time limit or deadline for our project to be completed. All that is expected is to accomplish "something" every calendar quarter. I know the day will come when the customer will say that he wants it now!

My problem is that I don't seem to be able to motivate my people to work at a faster pace. I have tried setting goals with each of them individually; I have tried letting them set their own goals; I have tried to set deadline dates for a particular part of the job to be done; but nothing seems to work. I have tried to work with them and actually help do their work. They seem to get their part of the job done when they get around to it.

Focus on the "do" not the "don't" items when confronting difficulty.

How should this project manager address this predicament? Figure 1-4 gives a few simple prescriptions that can help any project manager get started.

We also suggest considering the following five functions, as illustrated in Figure 1-5. The figure implies that the five functions are linked and interdependent. If the project is simple and straightforward, the functions probably will occur.

Do	Don't
Understand the difference between a task and a project, and understand that project management requires special tools because of the risk and complexity of the project.	Believe that software is going to do the thinking or manage the project for you. (For more, see Chapter 26).
Try to keep a big-picture perspective and focus on the integration needs.	Drive right into the tactical, and focus on the tools.
Pay attention to the needs (including the emotions) of people.	Avoid conflict.
Determine if you senior management has authorized this project.	Assume that things will work out on their own.
Remember that successful projects integrate the nine knowledge areas.	Focus exclusively on technical problem solving.

FIGURE 1-4. *Do's and don'ts for successful projects.*

1. *Defining.* Projects don't fail at the end; they fail at the beginning. Our project manager needs to develop some expectations and boundaries for the project. The project manager needs to start asking questions and determine the requirements for product performance and expectations for delivery date and cost. Good project objectives are SMART (*s*pecific, *m*easurable, *a*greed to, *r*ealistic, and *t*ime bound). Project definition also includes developing an understanding of management's rationale for investing in the project rationale.

2. *Planning.* The project manager and team need to examine their strategy for planning the project. This includes estimating resource needs, timing, budgeting, and so forth. The project members also should determine how they

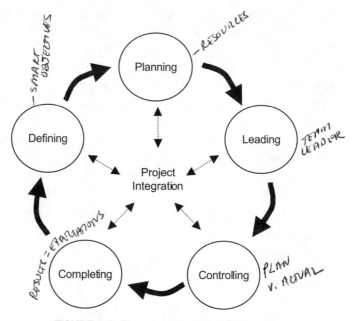

FIGURE 1-5. *Five management functions.*

will organize themselves to meet the customer's requirements and the control needs of their home (sponsoring) organization.

3. *Leading.* The project manager needs to provide managerial guidance to human resources, subordinates, and others (including subcontractors) that will result in their doing effective, timely work.

4. *Controlling.* The project manager needs to measure performance to find out how progress differs from plan in time to initiate corrective action. The controlling activities often cause a redefinition of the project objectives.

> **Projects don't fail at the end, they fail at the beginning.**

5. *Completing.* The project manager needs to make sure that the product scope conforms to the product requirements and the demands of the delivering organization (e.g., documentation and lessons learned).

This chapter establishes several fundamentals, including the definition of a project and of project management. As you continue reading this book, you probably will find yourself referring back to many of the framework ideas established in this first chapter. If you are starting a project—or in the middle of one—feel free to use the table of contents and index to help you identify the "just in time" knowledge and prescriptions that you need. We intend book this as a pragmatic resource.

HIGHLIGHTS

- Projects are a type of work distinguished from other types of work because projects are temporary, unique, and progressively elaborated.
- A task is similar to a project in that it is temporary, but typically, a project is larger and more complex than a task. Tasks are smaller elements of work in a project.
- A program is a collection of projects.
- Project management is not simply developing and managing a "to do" list.
- The factors that cause success and failure in projects are well known.
- Project management has a professional body of knowledge.
- This book presents five managerial functions for projects: defining, planning, leading, controlling, and completing.

Part 1

Defining the Goals
of a Project

2

Linking the Project to the Product

Customers and their product requirements are one determinant of project success. This chapter will help you to understand the relationship between the project life cycle to the product life cycle and give you other important knowledge that will help you get your project off to a good start.

Successful project managers manage expectations well. These expectations are set early in the project and reveal themselves in terms of needs, wants, requirements, requests, and other things. Good project managers continually manage expectations of the stakeholders. In this chapter, we will describe several important principles and rules for establishing and managing the goals and expectations of a project.

STRATEGIC ALIGNMENT OF PROJECTS

Recall from Chapter 1 that a project is part of a portfolio of assets. Organizations that perform projects consistently well do a good job of portfolio management. Projects are not an operational expense to be minimized but rather an important instrument for organizations to reach their goals. Top management has an important role in successful project management. First, top management has a fiduciary responsibility to invest the organization's assets wisely; thus, it commissions projects to align with its strategic intentions. Since all organizations have finite resources, management will select projects wisely to optimize value creation over the short and long term. Most organizations will find it better to select the projects that have the best returns on their investment. Second, top management recognizes that it must nurture and support its projects. The most innovative and

Projects are better considered as organizational investments than as operational expenses.

risky projects are often the most fragile because people are reluctant to have their names associated with a project that failed and stick with more conservative and conventional investments. Third, top management reprioritizes projects, or changes the boundaries and objectives, because the organization's environment is dynamic and things change. Project managers support top management by providing much of the data that allow them to make their decisions. Some of the goals of portfolio management include the following:

- Matching the number of projects for the resource base—such that the organization is neither spreading its resources too thin nor missing opportunities
- Setting and communicating priorities for creating value, balance, and strategic change
- Setting and communicating processes for escalating issues and managing change to project direction
- Setting and communicating completion criteria, including criteria for terminating projects
- Selectively leveraging the use of external resources such as consultants, contractors, suppliers, and partners

A project starts when it is authorized.   Thus, projects are investments of scarce organizational resources intended to achieve value-added organizational objectives. Here is one of the most important points in this chapter and indeed in the entire book: *A project starts when management authorizes it. Authorize* means that one or more responsible managers make a decision and communicate it to the rest of the organization.

The good news is that organizations (especially larger and better-managed ones) are growing in sophistication in linking portfolio management to project management. The bad news is that there are still thousands of organizations that do not understand the basics of project definition and project management, and that lack of understanding is putting their organizations' performance at risk.

Good project managers and project team members understand the relationship of their project to the goals of the organization. Mediocre project managers and teams simply comply and "take orders." You can help yourself and your organization if you think through and document the business case as early as possible. Strive to understand your project's place in the portfolio of all projects. Unfortunately, some organizations postpone or cancel projects too late rather than too early. This practice wastes resources and keeps the organization from exploiting opportunities. If your project has a weak business case, prepare to recommend its termination so that the organization can redeploy the resources on other projects. Help your organization make good decisions.

THE PROJECT LIFE CYCLE AND THE PRODUCT LIFE CYCLE

The product life cycle and the project life cycle are important concepts for defining projects. Generically, a life cycle frames the beginning and end of something. The *product life cycle* is the interval from the concept to the end of a product. The *project life cycle* is a part of the product life cycle, covering the interval from project initiation (also called *project authorization*) to project closure (or *termination*). Figure 2-1 illustrates the product life cycle and where one project life cycle could fit in. Understanding each of these life cycles and their relationships can help you to better define the start and finish of a project. In addition, you will gain a better understanding of how people measure product and project performance.

The product life cycle starts with an idea and ends with retirement of the product. Let's look at a software product as an example. Say that a sales representative for a software company calls on the human resources department of a large company and finds that there is a need for a new way of tracking employees' contributions to their pensions. This finding sparks the idea to develop an application for that client and sell it to them. The product is launched, and if it proves to provide value, it may (or may not) enjoy a long period of demand. Eventually, the undergirding product technologies become obsolete, and the company decides to no longer support the product; this marks the end of that product's life and the end of its life cycle.

An organization can commission a project at any point in the product life cycle, as is suggested by the examples in Figure 2-2. Thus, the project life cycle is a small part of the product life cycle. Furthermore, a single product life cycle can encompass several project life cycles. Compared with the product life cycle, the project life cycle has a more defined start and end point.

In addition to Figure 2-2, here are a few more examples that give further evidence that any given product life cycle can incorporate many types of projects. Product development could include installation of a brand new internal telephone system, a regulatory compliance program for newly enacted legislation, a new corporate Web site, or an advertising campaign. Product-life cycle management could include a project to modify credit card software so that only the last few digits are printed on the transaction slip (an enhancement) or an office rearrangement (a modification of an existing capability).

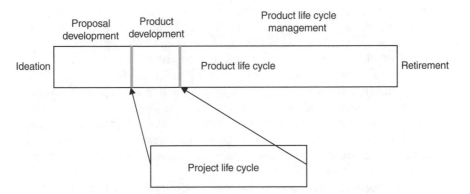

FIGURE 2-1. *The product life cycle and the project life cycle.*

Stage of the Product Life Cycle	Examples of Projects that Management Could Commission
Ideation	Research customer needs or create technologies that might have application for innovation.
Proposal development	Capture a specific business opportunity.
Product development	Develop and launch specific product concepts.
Product life cycle management	Enhance, upgrade, reposition a product.
Retirement	Withdraw a product from the market.

FIGURE 2-2. *Opportunities for projects within the product life cycle.*

Project life cycles are composed of phases, and phases are collections of activities that produce some kind of result. Often projects carry out phases sequentially, and this is termed the *waterfall project life-cycle model,* as shown in Figure 2-3. It is called the waterfall model because typically the project would complete one project phase before it "goes over the falls" to the next phase. The classic waterfall model typically has four to six phases. Some organizations mandate a life cycle (for the purposes of control and consistency). Other organizations give the project manager the authority to define the specific phases for the project based on the level of granularity of control desired by the project manager.

The project baseline is a culmination of tradeoff decisions on delivery time, cost, and requirements.

Note the event labeled baseline at the end of the design phase. The *baseline* is the culmination of the project planning efforts—bringing together all the nine project management knowledge areas described in Chapter 1. When a project manager declares that planning is complete, there is an implication that the project team understands its requirements, has made tradeoff analyses, and is ready to commit elements of the plan. In other words, the baseline is the approved project plan.

The purpose of baselining is to establish a series of comparison points for project control. Let's use a plane flight as an analogy of a project, with the flight crew being the project team. When an airplane takes off from Los Angeles headed toward Honolulu, the flight crew files a flight plan that describes the planned course they will follow (this is the baseline). Here the goal is clear-cut: Get to Honolulu. During the flight, the crew measures the plane's position, noting and reporting variances. The flight crew applies judgment to address those variances and takes corrective action to stay on course. If the flight crew were to make no flight-path corrections, the plane would land in the middle of the Pacific Ocean. Obviously, that is not desirable. The concept of a baseline is important to the management of projects, and we will stress it throughout this book.

From a project life-cycle perspective, one goal of project planning activities is to establish a realistic model of the project. The project Triple Constraint (described

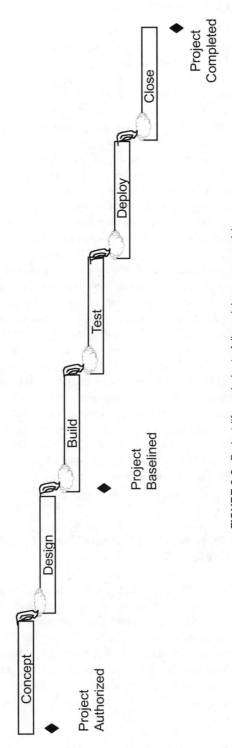

FIGURE 2-3. *Project life cycle (waterfall model as an example).*

in the next chapter) is a tool for conceptualizing and validating the tradeoffs for delivery time, cost, and product performance.

⚒ Better organizations have a "two contract" approach to initiating projects. They often call the first contract a *letter of intent,* by which management simply provides sufficient time and resources to the project manager and project team to develop a schedule and estimate for the project, typically including a risk analysis. The second contract is the *approved project plan,* where the scope of work, cost, time, and risk have been "baselined." The baseline is the point where the project transitions from "plan your work" to "work your plan." Baselines constitute another opportunity for management to approve the work. The benefit of the two-contract approach is speed and efficiency. Since it is difficult to develop good initial estimates, it is better to give the project team some early time to understand the requirements and validate its planning assumptions. This overcomes several dysfunctions that occur when people hold themselves or others up to an unrealistic standard of perfect estimates. One dysfunctional behavior is that people will shirk the project altogether or narrow their focus to their personal comfort zone. Another dysfunction is that some individuals will spend enormous amounts of time trying to get their estimate perfect. Finally, we note that cynicism and distrust often creep into the culture. The two-contract system tends to promote much more "ownership" and responsible behavior for the project.

Finally, we would be remiss if we did not point out that the waterfall model has some serious limitations. One major complaint about the waterfall model is that it emphasizes sequential processing of information and activities and thus is "slow." The complaint is well grounded in experience: Organizations that tend to rigorously adopt and use this project life cycle are those that place strong emphasis on product quality, consistency, and accounting. In other words, the waterfall project life cycle tends to reinforce bureaucratic controls at the expense of speed and agility. Given the competitive environment that many organizations face, a more dynamic and speed-based project life cycle needs to be used.

PROJECT COMPLETION INCLUDES DELIVERING A RESULT THAT MEETS THE REQUIREMENTS

Assume that initial requirements are incomplete and incorrect. Validate them by asking questions.

Requirements link the product life cycle and the project life cycle. Ideas that come from customers and users reflect their needs and wants. Customer may decide that they want to issue requests and solicit proposals from suppliers of project services. As the project progresses, customers continue to identify more needs and wants and make requests for additional functionality, performance, features, and services. These requests create both opportunities and problems for projects.

The term *requirements* is common in projects, but few people use consistent or generally accepted definitions. Thus, we start out with defining important terms:

- *Needs.* Something is a need if a system cannot function without it. For example, a person cannot function without food, water, and air. An internal combustion engine cannot function without fuel and air. A meeting room cannot serve its intended purpose if its temperature is intolerable.

- *Request.* A request is an expression from one party to another that invites action. A request is not the same thing as a requirement. A request is just that—a plea or directive for assistance in achieving a result.

- *Requirement.* A requirement is a capacity that a user needs to solve a problem or achieve an objective or a capability that a system must possess. A requirement is an agreement between the customers and users and the project "consuming side" and delivery side. Functional requirements describe what the user needs to accomplish or what the system needs to be able to do. Performance requirements are the quantitative dimensions that modify how well the function needs to perform.

- *Requirement specification.* This is a formal representation of a requirement. It is often a document, but it can be drawings or physical models. We provide an example below.

- *Wants.* People generally are motivated or take action based on their perceptions of what they want.

- *Statement of work.* In the professional lexicon of project management, the statement of work (SOW) is that portion of certain *formal* contracts [or proposed contracts in the case of requests for proposals (RFPs)] that describes the requested work. Many organizations use SOWs in a casual way to describe the product scope and requirements. The SOW is often done in lieu of documented requirements but contains explicit acceptance criteria and test specifications. Occasionally, the customer will issue an SOW that includes expectations for project schedule and cost. Regardless of whether an organization uses the narrow or broad definition of SOW, it is important to understand that the SOW is one way to communicate specific, measurable expectations for the project.

Here is an example of a product functional and performance requirement:

The brakes shall have the capability to stop a bicycle moving at 10 mph in 1.15 meters on dry concrete, under conditions of adhesion coefficient of 0.85 and rolling coefficient of 0.014. This performance shall be verified by company testing SOP # 23-08-02

A powerful question to ask early and often is, "What does done look like?" It is difficult to complete a project unless both the customer and the project team members can recognize and agree on the definition of project completion. Here is an example:

The Alpha project is complete when the following five conditions are met: (1) it has been at least six weeks after first shipment of product to customers, (2) the manufacturing quality assurance group has signed off that no significant issues are present in the product, (3) the sustaining engineering group has signed off that no more than 20 hours of effort are needed in support of Alpha over the coming 6 months, (4) the marketing product management has accepted ownership of the product support activities, and (5) project files are archived, including lessons learned.

Frequently, you will encounter requirements that are associated with the work effort rather than the result (or deliverables). Examples include the following:

- ISO 9000 documentation and compliance
- Environmental, safety, and health (ES&H) regulations
- Management policies for procurement practices (e.g., use standard parts or specific vendors)
- Special quality standards
- Corporate styling standards

Customers seldom hand the project a complete set of requirements specifications, and often the requirements they do identify are incorrect. One important reason for project failure is incomplete product concepts (some have called them "half-baked ideas"). Customers are often the source of these problems when they request solutions and specify the design (the "how") instead of their requirements (the "what"). One responsibility of all project participants is to recognize the issue, ask questions, and help the customer to specify the requirement. This means that early on in the project, the project manager and other project participants need to help the customer capture requirements.

Good requirements are unambiguous and only have one interpretation. For an example of an ambiguous word, *security* means different things to different people or in a different context. To a military contracting officer, it may mean a secret classification, whereas a person working on a military software project might interpret it as meaning data protection. At the Social Security Administration, the word may mean enough money to live well, whereas in the brokerage industry, it might mean stocks or bonds. Obviously, the intended meaning of any potentially ambiguous word must be understood clearly.

Sometimes the requirement suffers from vagueness, not ambiguity. For example, specifying a "good-quality finish" on a product's surface could be perceived as vague because different people could determine that the product result is unacceptable depending on different standards of "good," at least when working at the detailed level. The best way to deal with vague requirements is to determine how the customer will verify that the project has delivered an acceptable result. For example, you could verify that a food product was of acceptable quality if at least 8 of 10 participants on each of three randomly recruited consumer taste panels gave the new entrée a 6 or 7 (on a 7-point scale) for each of taste, aroma, and eye appeal.

Consider this riddle: In what way is managing a project similar to walking on water? It's easier if the product requirements specification and the water are frozen. In the ideal world, customers know exactly what they want and can express it. Thus, they freeze their requirements specifications, and the project proceeds in a linear way into design and deployment. However, in the real world, requirements specifications are seldom stable. Thus, the project manager and the project team need to remain alert for changes to requirements and evaluate those changes.

If requirements change, evaluate the impact on time, cost, and product performance.

Good project managers invest time to ensure that the requirements specifications are complete, correct, and understood. They talk with all key participants and then talk some more. Table 2-1 provides examples of some elements that may be part of a performance specification, and you want to be as certain as possible that you and your project team recognize and understand all these elements.

Clarify unclear requirements and design specifications.

In recent years, project management practitioners have recognized that it is both futile and wasteful to try to capture 100 percent of the requirements and to hold up project design work. In response, the project management profession has devised new approaches. For example, the practice of "timeboxing" is used frequently on information technology projects where a discrete period (typically 3 months) is used to develop and deploy a piece of functionality so that the customer receives value from the project quickly (rather than waiting for years).

THE DELIVERING ORGANIZATION AND THE CONSUMING ORGANIZATION

Projects that are performed internally in an organization face a common problem: confusion of the roles of the project and the customer. An example is an information

TABLE 2-1   Examples of Some Elements of a Performance Specification

Functional Features	Quality	Lifetime
For a car	Fit and finish	Shelf life
Top speed	Manufacturing yield	Mean time before failure
Number of doors	Defects per quantity of units sold	Details of customer support provided
Acceleration		Details of user service provided
Miles per gallon		Duration of spare parts availability
Trunk volume		
Service intervals		
For a digital camera		
Number of pixels		
Size		
Zoom ratio of lens		

technology department project that will deliver a management information system to another department of the same company. Because the project is internal, managers tend to focus on technical problem solving. Because managers have different values, they often perceive important integrative work as consuming much time and energy and providing little value. Consequently, we see common problems such as scope creep, lack of commitment, unnecessary conflict, and finger pointing. The remedy is straightforward: recognizing that projects involve agreements between two organizations, one of which we will call the *delivering* organization and the other the *consuming organization*. Figure 2-4 illustrates this relationship. Notice that the consuming organization acts as the requestor and funder of the project.

The first row of Figure 2-4 has some identified roles that people often confuse, and this role confusion compounds the problems. Use the following professionally

	Consuming Organization	Delivering Organization
Roles	Customer, buyer, funder, users Customer has a representative to manage details in accepting the product. This customer representative is also a project manager.	Developer, constructor Project manager, project participants, sponsor
Functions	Receives and uses product and its benefits	Develops and delivers the product
Role in Determining Requirements	Issues requests and participates in negotiations to determine requirements and changes	Receives requests and participates in negotiations to determine requirements and changes
Objectives	Wants the result to work, wants product quickly and for minimal cost. If there is no funding actually changing hands, customers tend to want everything. Has little incentive to constrain requests, especially when cost is no issue	Wants to satisfy requirements Has finite resources Some have a profit motive
Responsibilities	Product Life Cycle	Project Life Cycle
Escalation of Issues	The customer would complain directly to the project manager, and could have remedies of nonpayment or other contractual penalties	The project manager escalates issues to the sponsor.
Example of Escalation	The customer finds that product does not meet requirements. The customer's project manager would escalate the issue to the delivery project manager for resolution.	Project has insufficient resources or priority to meet objectives. Project escalates issues to sponsor for resolution.

FIGURE 2-4. The consuming organization and the delivering organization.

accepted definitions to reinforce the distinctions between the consuming side and the delivering organization:

- *Customer.* The person or persons who are *funding* the project. This person is also known as the *requestor* or *buyer.* The customer's definition of project success typically centers on getting a good product, at acceptable cost, and at the requested time. Sometimes the customer is the same person as the user; sometimes not.
- *User.* The person or persons who are using the product of the project. This person may be the *requestor* whose request initiated work or simply the *consumer.* The user's definition of success typically centers on product functionality (does the product do what it is supposed to do) and performance (does the product do what is supposed to do well enough.) Whereas a customer for an airplane might be an executive at an airline or leasing company, the users would include the flight deck crews, cabin crews, cleaning crews, ground crews, food service providers, maintenance technicians, and passengers.
- *Sponsor.* The senior manager (on the delivery side) who provides resources and visibility to the project. It is important to separate the role of the customer from the role of the project sponsor. The sponsor is part of the delivering organization, whereas the customer is the consuming organization. It might be helpful to remember that the word *sponsor* comes from the same Latin root word as *spouse:* one who provides for another. The confusion of the sponsor and customer causes many problems within projects, and it is our experience that good projects define the roles and responsibilities for the sponsor.

Let's examine deeper how each of the organizations will have different expectations, starting with the "consuming side" of the project. A single user of a product is very interested in reliability and ease of use and thus tends to ask for simple but robust product designs. However, most products have multiple users, and most users are trying to do something slightly different with the product. As a result, the functional and performance requirements for the product are such that only a complex design will satisfy all the users. As a rule, the more complex the product design, the more time and cost it takes to produce the design and the product. Not surprisingly, the customer (buyer) is interested in spending the minimal amount he or she can and wants delivery as fast as possible. The key point is that the customer cares about receiving the benefits. The project's efforts and frustrations are of little concern to the customer.

Customers have different expectations than the developers.

Now let's take a look at the expectations of stakeholders on the "delivery side" of the project. Project participants are often technical people whose personal mission in their work life is to gain and keep expertise in one narrow specialty. Their personal values and comfort zones often drive them to provide a design that is optimized around their personal specialty and maximizes that performance—

regardless of schedule or cost. In the case of fundamental research projects, the discovery of new knowledge or obtaining a patent is likely to be the hallmark of success.

ALL PROJECTS INVOLVE AGREEMENTS

In the preceding section, you read about a relationship consisting of a "consuming organization" and a "delivering organization." The consuming entity is receiving something of value from the project and, in return, should be paying for the value. If managers are going to be accountable for performance, they need to be able to judge the costs and benefits of their projects. In fact, we are seeing more and more companies develop project accounting and charging systems that allow them to evaluate the costs and benefits of projects. This exchange-of-value concept supports an important principle: All projects involve agreements. Good agreements protect both the consumer and the deliverer of project services.

Good project managers recognize and manage stakeholder expectations.

Organizations that perform projects consistently well always have some formal agreements in place. These agreements—for all practical purposes—are contracts and encourage accountability by all parties to the agreement. Recall from Chapter 1 that there is recognized knowledge area called *project procurement management*. Project management formality and rigor vary from organization to organization, and from project to project. For example, government agencies and departments conduct their projects with formal specific rules and contracts, whereas a homeowner hiring a contractor to install a swimming pool would have a much more informal approach. Some organizations allow a high degree of informally for their internal projects, and this informality often leads to mismatched expectations and poor project performance.

Since all projects involve agreements, we can derive a commonsense rule: You should only expect others to provide what they have agreed to provide and at an agreed level of performance. (These are simply the functional requirements and performance requirements that we described earlier.) A good agreement is not coercive, and responsible people will consider carefully whether they want to enter into an agreement. Thus, an individual is only obligated to do what he or she agrees to do. Consider this example: A project manager in the information technology department of a major transportation company was handed an assignment and told to "do it." The project had all the earmarks of a project destined to fail: an unrealistic due date, no organizational support, and immature technology. When the project manager asked for additional resources and time, he was told "no." This project manager then performed a courageous act: He resigned! His company's management was surprised but sought to understand his reasons. They saw that he was serious and that no amount of bullying and bluffing would cause him to agree to something that he felt was ill-advised. They were impressed by this show of principle and backbone and took the project manager's point of view seriously. Management's commitment to properly support and sponsor the project caused the project manager to reconsider his resignation and take on the project. The project was a success because it got the resources, attention, and leadership it needed. The project manager showed that he was an extraordinary individual is now a fast-rising executive.

The fact is that *we all have choices*. Unfortunately, many of our choices have unpleasant consequences. For individuals, the consequences include the wrath of powerful people in the organization. For organizations, the inability to choose and prioritize projects often results in death-march style projects where there is fear and intimidation, a diluting of strategic impact and profitability, and a culture that is compliant but unenergetic. Outstanding project organizations have the ability to develop and honor agreements between the project stakeholders whereas mediocre organizations avoid conflict, strive to please powerful interests, and hope for the best. Successful project management has more to do with developing, honoring, and enforcing agreements than it does with mechanistic tools and techniques.

> *Organizational climate and culture are the biggest determinants of project performance.*

GOOD BOUNDARIES

The American poet Robert Frost wrote, "Good fences make good neighbors." A good fence helps each neighbor responsibly steward his or her own resources without encroaching on others. In a similar sprit, good boundaries make good projects. One of the most important project management tasks is to partition the project so that everyone knows and agrees on what the project is going to provide and what it is not going to provide.

> *Managing agreements and expectations is all about setting good boundaries.*

We would like to introduce you to a powerful tool that will help you to establish project boundaries. In doing this boundary setting, you will be better able to set agreements and manage expectations. The tool is known as the *includes-excludes list* or the *in and out list*. It is rather simple to construct and is a two-column table with the word *Includes* at the top of the left column and *Excludes* at the top of the right column. Working together, the project team and the customer develop a list of requirements, product features, or work that is "in scope" and "out of scope" for the project. The customer then signs off on the "excludes" column. If the customer disagrees with an excluded item, he or she can negotiate for its inclusion on the "Includes" column. This approach also can help to clarify expectations for the phased product release common in software environments and other high-technology environments.

Do	Don't
Get a written authorization to start work on a project.	Assume that things will work out on their own.
Document requirements and assumptions.	Confuse a request for a requirement.
Recognize the difference between the product life cycle and the project life cycle.	Accept product support responsibilities as a *project manager* outside the scope of product and project requirements.
Ask questions.	Assume that just because a project is present that "someone, somewhere, knows what they are doing" in selecting and authorizing the project.

FIGURE 2-5. *Dos and don'ts for successful projects.*

TAKING ACTION

If you are a project manager with a new assignment, review some of the guidance listed in Figure 2-5. This is critical time for you and your organization: If you get the authorization step right, you are much more likely to be successful in your project.

In summary, here are some important insights that successful project manager know and apply:

- Projects are investments of resources.
- The user's or customer's wants and needs lead to the establishment of require-ments. Ideally, customers clearly and completely know what they want and need; thus, customers develop the product requirements specifications and pro-vide it to the project. However, in the real world, customers do not know as much as the project participants would like. Projects that actively collect and manage requirements are much more likely to be successful than those that do not. The job of the project is to deliver a product that satisfies requirements.
- Despite considerable *product* life-cycle activity such as product ideation, the project *only* starts when management of the delivering organization authorizes the project. They may need a contract with a customer before they commit.
- The project comes to an end when the project completes two steps: acceptance of the product by the customer and the "delivering organization's" administra-tive closure process.

Thus, management controls the start and end of a project.

HIGHLIGHTS

- Since projects are investments, organizations evaluate and select projects in light of their organizational strategy.
- Everyone on a project team should understand the rationale for the project.

- A request is not a requirement.
- Recognize two distinct organizations: consuming side and delivering side.
- Management authorization signals project initiation.
- Projects are complete when they meet their requirements *and* have completed the administrative processes of the delivering organization.
- All projects involve agreements.

3

Balancing Competing Demands with the Triple Constraint

The Triple Constraint helps the project team to evaluate expectations for product performance and compare them with expectations for delivery time and cost. Projects are work systems with multiple dimensions of performance, where the project manger balances competing demands.

The preceding chapter described three concepts that provide grounding for this chapter. We established that

- Managing expectations is an important project manager responsibility.
- Agreements between the customer and the project define what is in and out of the project. Requirements are one important kind of agreement that should be the starting point for project planning.
- The project baseline is a culmination of project planning and is the basis for project control.

In this chapter, we extend these points in describing successful project management as a process of balancing competing demands using the concept of the project Triple Constraint. We will start with a discussion of the multifaceted nature of project performance, introduce the Triple Constraint model, and conclude with a brief discussion of the decision-making process.

MANY WAYS TO MEASURE PROJECT PERFORMANCE

Almost everyone values time, and each of us occasionally gets impatient. Time is a visible part of our lives, and we focus much of our organizational life on the calendar. Timeliness is on most people's list of critical success factors for a project. In some instances, *time is the most important measure* of project performance. For example, consider the expectations placed on a large company's engineering department with a departmental mission of rebuilding industrial furnaces. For them, the *only* politically acceptable time available for rebuilding the furnace is during plant shutdown over the Christmas holiday period. For another example of the importance of timeliness, managers consider time to market important to their firms' competitive advantage. Finally, there may be significant penalties if a project is late. People use the term *deadline* in projects to create a sense of urgency rather than the more appropriate phrase *due date*. The perceived importance of time is inferred by this quote from one project manager. "The only thing that will get you fired here is to miss a date."

Of course, timeliness is not the only criterion for good project performance. A few additional performance parameters for the project and the product would include the following (recall from Chapter 1 that term *product* means any result—tangible or intangible—delivered by the project to the user):

- *Cost and price.* For many projects and organizations, the cost budget is more important than timing. If the project consumes more resources than the estimate, the overage will come from one or more of five sources: the profit margin, a price increase, people's personal time, reducing the product scope, or "deferring" work. We place the word *deferring* in quotations because many people put off immediate work but end up expending considerably more resources performing rework.

- *A product that satisfies its requirements.* We stressed the importance of good requirements in Chapter 2. It has been said that a customer will remember a poor-quality product long after he or she has forgotten that it was delivered on time. For example, the quality of the product is critical for medical product development because of moral, regulatory, and tort ramifications. Another example of the importance of subjugating cost and schedule performance to product performance is this: A science project had the goal to carry a large telescope by balloon to 80,000 feet, where it would be above almost all the earth's atmosphere, permitting the optical resolution of fine detail in distant nebulae. To accomplish this goal, the telescope's primary mirror had to be virtually perfect. Because the project could succeed only with such a mirror, there was no point in flying the balloon-borne telescope until this performance specification had been met. Thus, the project manager had to convince stakeholders to relax their expectations for time and cost performance.

- *Morale.* Some project managers use a heavy-handed, death-march style of leadership. We argue that if the project participants hate their involvement in the project, it would be hard to claim that the project is managed well.

There should be no doubt about the presence of multiple parameters of project performance. Furthermore, stakeholders develop and apply these parameters subjectively. A stakeholder's judgment of project performance, then, becomes a blending of things that went according to expectations, things that exceeded expectations, and things that fell below expectations. For example, customers are one important stakeholder. Customers generally want their results to be good, priced fairly, and delivered fast. The project manager needs to manage customers' expectations well; otherwise, their wishes might overcome the capability of the project. They might conclude that the project was a poor one, when in fact it was delivering to the requirements. The following section describes a helpful conceptual tool.

THE TRIPLE CONSTRAINT

The *Triple Constraint* is a project management term for a framework consisting of three parameters of project performance, commonly: product performance, the time schedule, and the cost budget. Figure 3-1 illustrates the Triple Constraint as three vectors:

- *Product performance*. The product performance "constraint" develops from the project team's capture of the product's functional and performance requirements (as described in Chapter 2). Note that money may appear in this parameter as factory (i.e., standard) cost of a manufactured product, the normal

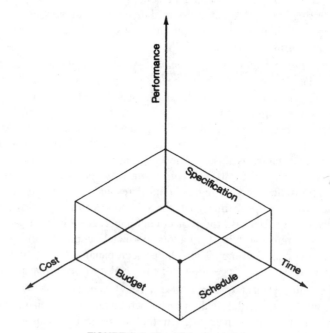

FIGURE 3-1. *The triple constraint.*

operating cost of whatever the project produces (e.g., miles per gallon for a car), the required maintenance cost of whatever the project produces, or some similar measure of product performance. (This money aspect of performance should not be confused with the project cost; see below.)

- *Time performance.* The project team determines the time side of the Triple Constraint (the project duration) by taking the list of activities, estimating their duration, and analyzing the critical path, as described in Chapter 7 and elsewhere.

- *Cost performance.* The estimated cost of the project, which may include capital expenses (even though these may be in the organization's capital budget as well), is computed through cost-estimating practices, as discussed in Chapter 9 and elsewhere.

Projects are work systems composed of multiple elements. In planning the project, the Triple Constraint guides some of the tradeoffs that the project team makes in developing the baseline. Our experience suggests that project managers mostly use the Triple Constraint as a planning tool. During execution of the project, the Triple Constraint helps the project manager to keep focus on the expectations that are typically important to the customer.

The Triple Constraint describes a relationship between product performance, time schedule, and money (or labor hour) budget.

Figure 3-2 illustrates the relationship between the three parameters. At a given level of performance, a given price curve will yield an estimate of the project duration. If you increase your budget, you might be able to find and apply more effective resources to the project, and this will shorten the schedule, as illustrated by the lower (dashed) curve.

Compared with Figure 3-1, which presents the Triple Constraint as a bounded three-dimensional space, there is an alternative view, which is a triangle. Figure 3-3 illustrates a view that makes the customer's preferences more visible. People who

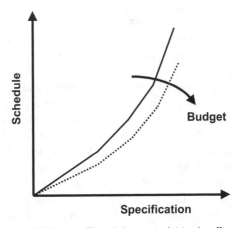

FIGURE 3-2. *The triple constraint trade-off.*

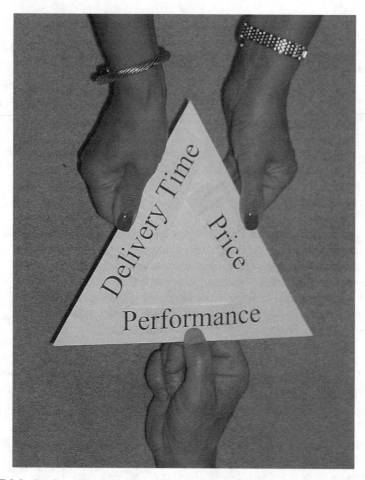

FIGURE 3-3. *An alternative depiction of the triple constraint. The customer can grab the two sides that are more important to her, and must accept the third side.*

use the triangle will label the sides with each of the parameters, although some people maintain that it is better to show the parameters on the corners of the triangle.

Here is a neat trick suggested by the way the hands are holding the triangle in Figure 3-3. The project manager has cut out a piece of cardboard and labeled the sides as shown. He has given the triangle to his customer (wearing the jewelry in this photograph) and asked her to select the two sides of the triangle that are her priorities. In this example, she has selected "Delivery Time" and "Price". She wants something that is fast and within her budget. The project manager is holding the third side, labeled "Performance," and he gets to determine the appropriate level of quality for his customer's indicated preferences. Try this on your next project; it just might help you to better set the expectations for the product and for the project.

A MODEL TO HELP EVALUATE COMPETING DEMANDS

The process of project planning is a process of constructing a project model. You can model any system, and the model can be simple or complex. The Triple Constraint is a simple three-element model of a project system. It is easy to understand because there are only three relationships between the elements: product performance and time, time and cost, and cost and product performance.

Now, let's see how project managers apply the project Triple Constraint Model. Figure 3-4 illustrates relationships between the three vectors of the Triple Constraint and shows time as the independent variable and cost as the dependent variable. In other words, for a specified cost level and with a given Triple Constraint relationship, the budget can be determined. However, you needn't see the relationship as "mathematically" as it is depicted in Figure 3-4. You could view the product performance specification as variable (as suggested by the arrows showing a better direction when the curve moves to the right or a worse direction when the curve moves to the left). Alternatively, you could specify both time and cost, and the product performance would be determined by those constraints.

Here is the important insight that the Triple Constraint brings: In a simple system of three parameters, two parameters can operate independently, and the third parameter is dependent. Since customers (and their requirements) always should be considered, the model helps the project to probe for customer preferences, namely:

- The customer can constrain one side.
- The customer can optimize the second side.
- The customer needs to accept the results of the third side. It "free floats."

Consider this example to help you understand how project managers use the Triple Constraint Model to manage customer expectations. A developer requests a

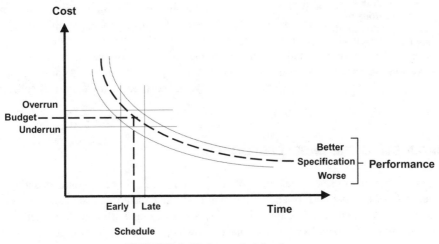

FIGURE 3-4. Triple constraint outcomes.

graphics artist to design a high-quality logo for a shopping mall. However, the mall developer's budget contains only a small amount of money for the artwork. Unfortunately, the shopping mall developer has not involved the artist in project planning and is now approaching an important due date. The artist replies, "You can have it good, fast, or cheap. Pick which two you want." While she might have been a bit blunt, she provides a good example of how understanding the Triple Constraint can help in managing expectations. The simplicity of this three-element model makes the choices more visible to the customer. It infers that the customer may have to accept less-than-ideal outcomes. There are also important implications here in that a good project manager (the mall developer) will involve project participants (the graphics artist) early.

Simple models help people to understand tradeoffs. However, recognize another important point in model building: The simpler the model, the lower is the fidelity. A model that has higher fidelity is more realistic and has more power to explain and predict. Based on our discussion thus far in this book, you should be able to see that it is overly simplistic to state that good project performance and success are a matter of being "on time, according to spec, and within budget." We frequently hear project management "experts" who argue that the Triple Constraint has four (or more) sides! They point out (correctly) that such things as project risk and team morale can affect outcomes positively or negatively. Therefore, they add these elements to their model to make the model more realistic. Certainly, their models are more robust, but they are also more complex (and difficult for the layperson to comprehend). For example, a model of four parameters will have six tradeoff relationships, and a model of five parameters will have ten tradeoff relationships. Many people do not have the patience or motivation to want to understand project management at the level of sophistication that many project management authorities want them to.

> **The project manager has to be able understand the complexities of a project but communicate them in a simple way.**

There is an important managerial implication of our discussion of modeling: As a project manager, you need to understand the nuances and complexities of *your* project models. You have to be able to grasp the details as well as the big picture. You expect your model to be able to help you to anticipate the future of your project. On the other hand, when communicating and negotiating, you have to identify and select the most powerful constructs and explain the relationships between the model elements in a simple way.

ADJUSTING THE BASELINE FOR RISK

It is an important project management responsibility to help people understand the tradeoffs of product performance, time, and cost. As part of this responsibility, you need to understand how risk analysis and response planning affect the expectations. It is common for people who don't have knowledge or experience in a project to assume that the project will be quick and easy. Be skeptical of those claims, and consider involving expert technical consultants to help in the risk analysis.

As we will describe in Chapter 11, a risk is a discrete occurrence that may or may not happen, and if it happens, it may affect the project positively or negatively. Most good project managers will perform risk identification, analysis, and response planning during the planning process. As part of a quantitative analysis of the risk, they estimate the impacts on all dimensions of project performance: requirements risk, scope risk, cost risk, schedule risk, and so on.

It is the rare project that does not have an occurrence of a risk event. One way to manage these risk events is to create a risk management reserve. Project management professionals also know of this reserve as a *contingency* or a *buffer.* Good projects build this reserve right into the baseline, and the agreement for the project acknowledges (rather than denies) the presence of risk and holds the project accountable for managing project risks. Thus, the project planning process includes *adjusting* the baseline for risk.

We acknowledge that *many organizations do not adjust their baselines for risk.* Some people and organizations are simply ignorant of basic practices, and others have cultures that criminalize risk. This is regrettable, but it is also reality. Our intent is to highlight the practices of *good* project management. It has been our experience that when the project team, sponsors, and customers all believe that the baseline is valid, they will be more likely to agree that the project is a success.

We also acknowledge that many project managers feel compelled to accept an unrealistic schedule or budget for a project when some powerful party (such as the customer or the boss) dictates all three sides of the model. In other words, the project model is invalid and unrealistic. The fact that this happens does not excuse it. The solution? The project manager needs to conduct a risk analysis, be creative, build coalitions with other power bases, and practice good leadership.

HOW THE TRIPLE CONSTRAINT HELPS TO EXPLAIN THREE COMMON TRADEOFFS

The Triple Constraint framework explains much about tradeoff decisions made in projects. The following paragraphs describe the tradeoffs involved in three project management practices: resource leveling, rightsizing the product, and crashing.

Let us look at how project managers balance competing demands. In each of the three examples, we are showing how one dimension of the Triple Constraint is constrained, the second dimension is optimized, and the third is the dependent variable.

Resource Leveling

Resource leveling is the project management term that describes adjustment of the schedule to accommodate resource considerations. Resource leveling is generally a decision made during the planning activities, before establishing the baseline. For example, if limited resources were available, we would expect the project planning process to extend the duration of tasks or change the way the schedule's float (also called *slack*) is allocated in the schedule. When resources are "leveled," we would expect the project to take longer. In terms of the Triple Constraint, the project is holding product performance (requirements) constant and is trying to optimize resources. What does the project give up? Schedule.

Rightsizing the Product

Ideally, project planning should start with requirements capture and specification, and this determines the product performance side of the project Triple Constraint. However, it is a common occurrence for organizations to start projects with a date assignment, and decision makers within the project fit the work scope to the available resources. Thus, the requirements satisfied are a compromise between the available time and the available resources.

When projects *rightsize* the product, they are adjusting the amount of project performance for the available resources and available schedule. For example, in delivering samples of products to a trade show, it is vital to be ready by the show date, even if it means that the samples lack full product functionality. In terms of the Triple Constraint, the project usually is holding schedule constant and is trying to optimize the resources. What does the project give up? Product performance (requirements).

Rightsizing is a decision that is and should be made during the planning activities as a way of prioritizing requirements. In the real world, customers are unclear about their needs and requirements, and technical people always would like more detail. Thus, the reality for many projects is making rightsizing decisions during execution.

Crashing

Crashing is a project management term that means to spend more money on the project in order to speed up accomplishment of scheduled activities. Crashing is a decision made during the execution and control activities and not in planning activities. In some cases, it is possible to apply better-quality talent resources to the project. Assuming that these resources are more expensive and productive, the project accomplishes its work faster. What does the project give up? Cost.

> **The project manager can "crash" the schedule by using more productive resources, which normally are more costly.**

In terms of the Triple Constraint, the project is holding product performance (requirements) constant and is trying to optimize schedule. It is willing to spend more to achieve the result. Note that sometimes you can overapply cost and resources to a project and actually reverse the time improvements (in other words, the extra resources cause additional complexity that causes inefficiency that slows down the project). The project manager needs to analyze carefully the tradeoffs before committing to the crashing technique.

Crashing is a decision made during the executing and controlling activities, after the project is baselined.

THE TRIPLE CONSTRAINT DURING CONTROL

Recall from Chapter 2 that an output of project planning is a *baseline*. Thus, the Triple Constraint concept of a baseline implies that the project team has done the necessary planning activities to understand

- The work needed to meet product performance requirements
- The time needed to accomplish the work
- The resources needed to accomplish the work
- The risks involved

The presence of a baseline also implies that top management has agreed to the assumptions in the project model. This baseline event signifies the end of major planning and activities and the start of executing and controlling activities. After the project team has established and received approval for its baseline (it has "planned its work"), it now must "work its plan." The project then works to the baseline plan, monitors against it, and takes corrective action when required.

In the paragraphs that follow, we describe two common problems that you may face during execution of the project: scope creep and gold plating.

Scope Creep

The Triple Constraint concept will help people to understand a common project problem known as *scope creep*. It is normal for a customer to make requests for additional features and functionality in a product. We define *scope creep* as adding work scope *without* adding resources and time. If you agree to provide these extra features, which were not part of your project's baseline, you are agreeing to do additional work (perhaps even modifying the design)

Scope creep is a failure of the project manager to apply basic change management concepts.

without changing the schedule (or cost). Scope creep often leads to schedule delays and budget overruns. In a large project, any single change probably only has a trivial impact on the schedule or budget. However, the cumulative effect of many small changes can be significant. It does not take many changes of this sort first to produce a one-day schedule slippage and then a one-week slippage and so on until the project is in serious difficulty. Wise project managers only agree to undertake additional work after analyzing the proposed change's effect on the project baseline. If a change is accepted, you must modify the contract or other documents that define the project's goal, schedule, and cost, as discussed in Chapter 20.

Gold Plating

Overachieving the product performance baseline is often called *gold plating,* a jargon term that means adding additional performance and cost to the product that exceed the customer's requirements. Many engineers, scientists, and other technical experts have individual personal values that overemphasize their subject matter expertise (e.g., elegant design or new scientific knowledge) and underemphasize other goals (e.g., schedule or cost). Gold plating is typically a combination of these personal values combined with the project manager's inability to influence their behavior. We suggest the following insights:

Keep project team members focused on meeting requirements, and avoid gold plating.

- "Better" is the enemy of "good enough."
- "Perfection" is not the same thing as "meeting requirements."
- The project manager is responsible for setting expectations and directing the work in order to meet requirements on schedule for the allocated budget.

OTHER EXAMPLES OF BALANCING COMPETING DEMANDS: FINANCIAL MANAGEMENT

In many organizations, the project manager has to consider the source of funds and project profitability. Accordingly, the project manager has to make tradeoff decisions and negotiate for the outcomes that will make the project successful. The following paragraphs describe how one might consider balancing competing demands for a cash-flow shortfall and for a project profitability squeeze.

Cash Flow

On occasion, the customer may not fund the project as expected, and this may cause a deficiency in cash flow. (Cash flow is not the same thing as project performance management, such as that described by the earned-value technique in Chapter 19.) Suppose that your project has a duration of three months, and the customer originally proposes to fund you with $100,000 per month (i.e., the customer will fund your project via progress payments). The first month has gone according to plan, and you have received your $100,000. At the start of the second month, your customer informs you he has only $50,000 available for that month but that he will have $150,000 available the third month. You will still get $300,000, but you cannot apply the planned resources during the second month. To catch up to the schedule baseline, you will have to "crash" the project's schedule by spending more money during the final month. When confronted with this sort of re-funding proposal by a customer, therefore, the prudent project manager will evaluate options for cash flow. Perhaps the customer can be convinced to provide the cash flow, or the sponsor can provide it, or partners, or the financial markets. Of course, the project can eliminate some requirements or extend the schedule.

Profit Adjustments

Sometimes the decisions made during competitive negotiation affect the baseline. Imagine that a contractor has bid $10 million to build a shopping center and is very motivated to win the work. During negotiations, the contractor was told that unless the price is lowered to $9.5 million, the contract will be awarded to another company. The contractor agrees to minor wording changes that "appear" to reduce the scope of work a bit and to justify a substantial cost reduction. With this change, the probability of a cost overrun has increased. A good project manager will understand how much of that $500,000 reduction originates from the following:

- The reduction of planned profit
- Planned efficiencies in producing work
- The elimination of requirements

In this example, we also need to note that many contractors will follow a "low balling" strategy, meaning they will deliberately underprice the work with the expectation that they will recapture any foregone cost and profitability in contract changes. If you are the project manager for a project owner, you need to be aware of this practice and include its potential in your risk management planning.

PROJECT MANAGEMENT AS A DECISION-MAKING PROCESS

To restate an earlier assertion, successful project management is a process of balancing competing demands. Good project managers realize that you can't just focus on one performance parameter to the exclusion of others. The customer can get fast delivery, but the product may be expensive or not perform well. The customer may get a great product but wait a long time for it. The price may be cheap, but the product may not be very good and the delivery time long. Customers—as well as other stakeholders—often will raise their expectations without informing others.

Over the years, we have worked with thousands of project managers and seen some very good ones and some very poor ones. The better ones understand the systematic nature of the project and the importance of balancing competing demands. The less competent ones focus on one dimension

The project is a system that should be optimized.

(such as technical performance) until something (such as the schedule or performance) becomes a crisis. The better project managers know that successful project management involves treating the project as a system of people, capital, energy, and so on. As a system, if one element changes, then there is an effect on other elements. The manager makes tradeoff decisions among a number of project and product variables to produce an optimal result. The essence of project management is decision making.

HIGHLIGHTS

- There are many ways to measure project performance.
- Project planning is the process of developing a model of a project. Modeling is a process of trading off simplicity for fidelity.
- The Triple Constraint helps to evaluate planning assumptions and tradeoffs in developing planning for the project baseline.
- The Triple Constraint helps the project and customer understand the process of balancing competing demands. Examples of these competing demands include resource leveling, crashing, rightsizing the product, scope creep, and gold plating.

4

Contracts, Negotiations, and Proposals

This chapter describes the process of developing and formalizing agreements between the project and the customer. Formal contracts are one important means of showing this agreement, and contracts apply to both internal and external projects. Many contracts are the result of a proposal and negotiation process.

Part 1 of this book focuses on the importance and methods of defining a project. In the preceding chapters, we described the distinguishing characteristics of projects and how customer requirements establish the basis for measuring success. We established the principle that all projects involve agreements. We described how the project and the customer arrive at those agreements through a process of balancing competing demands.

In this chapter, we wrap up our discussion of project definition with a discussion of contracts, negotiations, and proposals.

CONTRACTS

In Chapter 2, we noted that all projects involve agreements and provided examples in the areas of requirements, management's authorization, and baselining. The word *agreement* is really a softer word for the more legalistic term *contract*. Thus, when a project manager, team, or sponsor commits to delivering something to a customer, that commitment is a contract. It does not matter if the project is internal or external, the promises made constitute a contract.

> **By documenting your agreements and intentions, you will avoid future misunderstandings.**

Admittedly, people perceive internal projects differently, particularly the extra work needed to document and negotiate their understandings. People may not perceive this extra work as beneficial, but if the project performs poorly, almost always people point back to poor "contracting" as a cause of the poor performance.

> **Knowledge of contracts is of increasing importance to project managers.**

You also should recognize that one of the biggest trends in project management over the past decade is the use of partnerships, codevelopment, outsourcing, and other combinations of organizations. It is vitally important that project managers understand the rudiments of contracts.

A valid contract must meet the following three tests:

- There is an offer by one party to provide something to another party. It must be for a legal purpose. For example, if two drug dealers agree to sell one another cocaine, a court would not uphold the agreement because selling cocaine is illegal.
- There is acceptance of the offer by the other party. Often negotiation is used to balance the offer and acceptance. Each party voluntarily enters into the contract (no coercion is involved).
- There is "consideration" involved, which means that something of value (such as money) is exchanged.

> **Formalized agreements have less potential for misunderstanding.**

Courts of law, including the U.S. Supreme Court, recognize that verbal agreements are equally binding and enforceable as are written agreements. Nevertheless, it is a good idea and common practice to formalize agreements into written contracts.

> **Document your agreements.**

Figure 4-1 shows that two parties are involved and that the contract bridges the understanding between the two organizations.

There is a variety of possible contractual forms (Table 4-1). In the first of these, very common in commercial situations, the contract is called a *fixed-price* (FP) or *firm-fixed-price* (FFP) *contract.* This has the lowest financial risk to the customer because the maximum financial obligation is specified; conversely, the FP form has the highest financial risk for the contracting organization but offers the highest potential reward if the estimated costs can be underrun. This is illustrated in Figure 4-2. There are variations on this in which the price allows for escalation or redetermination owing to some set of factors, such as inflation. Alternatively, there may be an FP contract with an incentive fee based on some performance aspect, perhaps early delivery.

Cost-reimbursable agreements are another type of contract. Here, the customer bears an obligation to reimburse the contractor for all costs incurred, so the customer has a high financial exposure, and the contractor has a correspondingly low

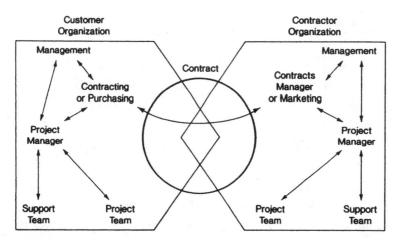

FIGURE 4-1. *Interpretation of the triple constraint must be controlled through the contract.*

risk. Typical contracts of this sort are *cost plus a fixed fee* (CPFF) and *cost plus an incentive fee* (CPIF). *Time-and-material* (T&M) and *rate-of-effort contracts* are a form of cost-reimbursable contract.

An FP contract is preferable when there is high confidence in your ability to satisfy the performance specifications on schedule and for your cost and profit targets. Such situations typically occur when you have done essentially identical or very similar project work previously, and you have the appropriate human and physical resources available. If these conditions are not satisfied, the FP contract is a financial gamble and should be undertaken only if the risk is acceptable and the prospective reward is commensurate with it.

Contracts entered into with the U.S. Department of Defense are governed by complex regulations that are special versions of federal procurement regulations. In the case of FP contracts, the government is never obligated to pay more than the specified amount. However, if the contractor performs very well and manages

TABLE 4-1   Common Contractual Forms

Abbreviation	Definition
FFP	Firm fixed price—the price and fee are predetermined and do not depend on cost
FP	Fixed price—same as FFP
CPFF	Cost plus fixed fee—the customer agrees to reimburse the contractor's actual costs, regardless of amount, and pay a negotiated fixed fee independent of the actual costs
CPIF	Cost plus incentive fee—similar to CPFF except that the fee is not preset or fixed but rather depends on some specified incentive
T&M	Time and material—the customer agrees to pay the contractor for all time and material used on the project, including a fee as a percentage of all project costs

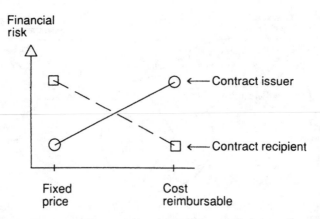

FIGURE 4-2. *The contract issuer and recipient have complementary financial risks.*

to underrun the cost budget substantially, the government has the right to reduce the amount paid to the contractor below the price specified in the contract. Thus, FP contracts are a one-way street to the government's advantage.

Legal Aspects

There is a myriad of legal aspects in project contracts. Discussions with suitable people in your organization are often helpful.

In the case of U.S. government contracts, there is a host of special regulations. A typical government contract may include the following:

1. Customer's and contractor's name and address and specification of authorized signatures
2. Statement of supplies (items), services, and prices (statement of work)
3. Preservation, packaging, and packing instructions
4. Delivery or performance period
5. Inspection and acceptance terms
6. Contract administration data
7. Special provisions (funding limitations or customer-furnished equipment)
8. General provisions (reference to federal procurement regulations, overtime payment terms, or similar) such that, for example, manufacturers have been obligated to pay criminal and civil penalties for violations of federal regulations
9. Patent terms, conditions, and ownership rights (if not covered in items 7 and 8)
10. List of required documentation

Despite all this intricate detail, it is still possible to find ambiguity. One company sued to invalidate the government's right to contract for delivery of systems for which specific congressional payment authority was not yet enacted. Although this is largely a contractor's ploy to renegotiate for a higher contract price, it demonstrates how seemingly clear contracts can be disputed.

Misrepresentation of costs is illegal and can produce serious consequences. There can be other problems with government contracts. There are many instances when government contractors (federal, state, or municipal) are vehemently criticized for allegedly inadequate project performance, and this is often widely reported in the media. In addition, in the case of FP contracts, failure to deliver can expose the contractor to very serious cost penalties far beyond absorbing the cost of its own effort. Such penalties not only may include termination but also may obligate the contractor to pay the costs of a substitute the government hires to deliver after the contract is in default.

If any patents are obtained in the course of the project, the government may own them. In addition, there are frequently requirements specifying how much of the work throughout the project must be performed. Costs frequently are subjected to audit when the job is completed, and the amount finally paid to the contractor frequently is reduced.

In construction projects, for example, there may be clauses about what will be done if unanticipated rocks are encountered below ground, delays are caused by unusual weather, or similar situations occur. *Force majeure* (i.e., an act of God) may permit a contractor to escape any penalty for failure to deliver on time. In addition, the suitably drawn contract will specify that disputes are to be settled by arbitration rather than by the courts. Arbitration is far simpler, usually quicker, and sometimes less costly to both parties. Nevertheless, antitrust regulations and many other laws limit the kinds of commercial arrangements into which two companies may enter.

NEGOTIATING THE CONTRACT

Contract negotiation really begins in the proposal phase because the expected contract form must be consistent with the job to be undertaken. If there is any reason to believe that the customer will require an FP contract (Table 4-1), for instance, and the job calls for a major technological advance you are not certain you can achieve, it would not be prudent to continue in the preproposal and proposal ef-

Early on, learn about the terms and conditions of the proposed contract.

fort. Hence, one objective of preproposal activity and discussions with a customer prior to major proposal expense is to ensure that the contract form they intend to issue is consistent with the contract form your company or organization is willing to negotiate considering the work to be undertaken. In addition, negotiations are designed further to improve the likelihood of the customer organization and contracting organization having the same perception of the job.

If the preproposal process did not remove these potential misperceptions, the negotiation process offers the last opportunity to do so. Specifically, the final deliverables—oral, written, or a more tangible product—must be well defined, and the criteria for measuring or judging acceptance and completion must be straightforward. In some instances, acceptance occurs only after the product has left the manufacturer's possession, such as, for instance, when a satellite must perform to some specified level in orbit or the actual operating cost of a piece of machinery must be less than a stipulated amount when that machinery is operating in a factory.

In a typical negotiation, the customer attempts to increase the performance specifications while reducing the schedule and budgeted cost. If it is a competitive solicitation, the customer often will play off one prospective contractor against another to try to maximize his or her apparent benefit. A common negotiation problem is deciding how to cope with the customer's request to lower the price. Several things help:

1. A good plan, well explained
2. A clear understanding of your risk-response planning (Do you know if anyone has inserted "padding," "cushion," or "fat" in their estimates? As Chapter 11 will explain, this is distinct from contingency budgets developed as part of the risk response plan.)
3. Management guidance, or clearance, on how much you can give up
4. A reputation for having met prior commitments

In addition, negotiators should understand clearly how far, if at all, their management is prepared to deviate from the terms and conditions offered in the submitted proposal it approved. Second, good planning aids negotiators in that there is a complete work breakdown structure with an attendant activity schedule and cost estimate for each element of it. Project management software can be helpful in planning "what if" tradeoffs. These help negotiators to understand the job being negotiated and usually can help to explain and/or defend it to the prospective customer.

Develop a negotiating strategy.

In preparing for negotiations, it is frequently desirable for your organization to conduct a trial run with someone or some group playing the customer. In short, be well prepared and know your minimum acceptable position. Also, because only one negotiating team member should talk at once, you can rehearse who will respond to particular issues if they are raised.

During actual negotiations, it is usually to your advantage first to define the job (the detailed statement of work, specifications, and test criteria) and the schedule. After that, you can negotiate the exact contract form, including any detailed terms and conditions and the final price. The effective negotiator always "horse trades" and never makes a unilateral concession. Nevertheless, there is often give and take in the negotiation process, and some changes may be agreed to. Whenever there is a change in one element of the Triple Constraint, there must be changes in other elements. For instance, a customer may offer to provide customer-furnished equipment (CFE) to reduce the expected cost of some activity within the proposed project. In the case of the government, this is called *government-furnished equipment* (GFE). When this occurs, language on the performance axis must be changed to indicate the performance specifications the CFE must meet, and the schedule of CFE provision must be stipulated. When both these things have been accomplished, it is possible to agree to substitute the CFE for contractor-procured items and to offer a reduction in the proposed schedule or budget. There are many possible tradeoffs, including, for instance, deciding who is responsible for laboratory space and equipment, physical security, test equipment, parts storage and control, and so on. If you are asked to surrender schedule or price, a counteroffer is to alter the performance specifications or obtain customer-furnished equipment or services.

Only contract-change notices suitably signed and agreed to by both parties can permit changes. Such changes should not occur outside the contracting mechanism, despite agreements reached by members of the contractor's and customer's support teams—for instance, when two engineers meet and agree that some new feature would be a desirable

Contracts are an important element of project change management.

item to include. Avoid verbal redirection. Renegotiations such as these require the same amount of planning and preparation as the original negotiations. It is not uncommon to have several renegotiations during a long or complex project.

PROPOSALS: A SPECIAL KIND OF PROJECT

Recall from the discussion of the product life cycle in Figures 2-1 and 2-2 that projects can arise at any point in time in the product life cycle. Thus, preparing proposals is a special kind of project.

Figure 4-3 illustrates the strategic framework for obtaining winning projects. It does not matter whether the projects originate outside the organization by a customer or within the organization. However, many organizations dissipate their energies in preparing losing proposals that

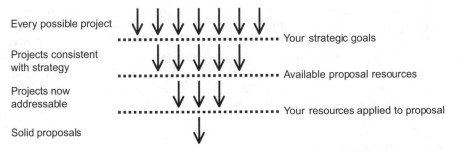

Every possible project — Your strategic goals

Projects consistent with strategy — Available proposal resources

Projects now addressable — Your resources applied to proposal

Solid proposals

FIGURE 4-3. Systematic filters are required to focus your efforts to obtain winning projects.

1. Cannot win the competition
2. Understates the feasibility and risks of delivering a solution that meets the requirements within the expected time frame and cost budget (All organizations have finite resources, and the more successful companies concentrate these on the few most attractive projects. To do otherwise starves some or all key efforts for critical capabilities. The goal of this section is to provide you with some orientation to a few of the concerns that confront senior managers in considering which efforts to pursue or authorize.)
3. Can be successful but are insignificant or irrelevant to the proposing organization

Even with the filtering illustrated in Figure 4-3, you never know which specific proposal you will win (if they are written for an outside customer) or which will be approved (if they are for a customer within your organization). Thus, because you can never be certain which project proposals will be approved, the organization is always adjusting resources and workloads.

> **Proposal development can consume considerable resources.**

Therefore, the basis of a successful strategy is to filter out losing projects. These include projects inconsistent with the organization's long-term goals or with the current and near-term resources within or otherwise available to the organization. Such filters might reject consumer project efforts in an industrial product company. Similarly, the filters could reject a fixed price contract for a technical development, scientific advance, invention, or other breakthrough in a conservative company that normally will not undertake a fixed-price contract for something not previously accomplished.

Filtering out the huge number of possible projects the organization might have addressed leaves a much smaller number of projects that are appropriate for the organization to consider. It can then address some or perhaps all of these in proposal efforts to which it applies adequate and appropriate resources. The result of this process is that an organization submits only well-founded proposals for consideration. This result is most likely to be achieved if the organization has a careful review process, often called the _bid/no-bid decision_.

In summary, avoid projects that are inconsistent with your organization's long-term goals or current and near-term resources, unlikely to win the proposal competition, unlikely to satisfy the requirements, or insignificant or irrelevant. There is a special situation in which your immediate superior demands that you undertake a new effort but you and your department or group are overloaded. In this case, your response must indicate the nature of the problem, suggest some alternatives, and indicate how you plan to satisfy this new requirement.

Bid/No-Bid Decision

The decision whether to bid on a proposal opportunity, whether to an external organization or within the organization itself, must be taken within the context of the organization's strategic framework. This framework, of course, is specific to the organization at that particular time. The company that rejects consumer product projects today may have the interest and capability to undertake such projects five years from now. There are many issues involved in this decision, four of which we discuss.

Funding   The first issue is whether the customer's organization has made a commitment to fund a project. This implies that the customer has done some work to define its requirements and develop a business case. All organizations have wants and needs but not enough money to satisfy those needs. For instance, many isolated rural villages in undeveloped nations could use an electrical power generation system, but these villages and the nations

> **Make sure the customer has the money to spend.**

of which they are a part cannot afford to pay for their installation. Contractors occasionally seek subcontractor proposals for "window dressing" to satisfy someone or otherwise justify an alternative approach they propose to take. Clearly, it is not worth responding to such requests for a proposal because the effort will not result in a winning project.

Proposals to external organizations may be written in response to a request for quotation (RFQ) or request for proposal (RFP) or because you have identified an opportunity. Business you may choose to solicit from external sources is illustrated in Figure 4-4, and the process is called the *proposal funnel* because you must keep injecting inputs at the top (i.e., find more prospects) if you expect to get anything out of the bottom (i.e., get more self-initiated projects). Again, not every possible project gets initiated, and there is no assurance that any amount of preproposal work will result in a specific project. W is a larger number than X, which is larger than Y, which is larger than Z, which is normally larger than 1.

It is also important to examine the priority or importance of the proposed project. In this connection, the relationship of a particular proposal opportunity to present and future programs is also an important issue. Many small "paper study" projects, not attractive per se, actually are very attractive because they can lead in the future to large production programs.

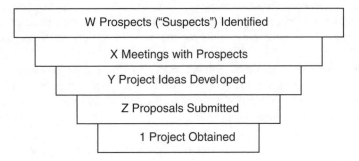

FIGURE 4-4. The "funnel" of effort leading to a solicited proposal. A measure of proposal efficiency is the size of the numbers Z, Y, X, and W.

Value Across the Product Life Cycle   The project's social, ecological, and energy impacts might be highly significant to an organization in deciding whether or not to bid on a project. Presumably, a project to build the prototype of a device to clean up oil spills economically would be more attractive than a project that endangers the earth's ozone layer. Similarly, a project that offers the opportunity to apply important new technology or otherwise enhance the organization's reputation might have high nonmonetary value. Conversely, the expected sales of a commercial new product project effort and the profits of such an effort also would be significant issues.

Finally, many projects are merely a hidden obligation for an organization to accept future financial commitments for new capital or facility investments. This must be discovered before any money is spent so as to be certain that future major financial commitments are within the organization's resources and abilities, as well as being consistent with the project's prospective value.

Response Ability   The central issue here, also illustrated in Figure 4-3, is the organization's present capability first to prepare a winning proposal and second to perform the proposed work. If some capabilities are not actually present, there must be a viable plan to make them available when they are needed.

Winning the Competition   First, an organization must ask whether there is advance information about the project available to it. This is particularly true about efforts arising from a customer organization and being presented to a contracting company as a request for a proposal (RFP), but it is also relevant for efforts within an organization. Lack of advance information often indicates that someone else has a head start or that the request was created hastily and lacks substance.

A second issue concerns the customer. Is it an individual (yourself, a friend, or someone else) or a commercial organization? If the latter, is it your organization or an external organization? If it is your own company, has your superior ordered you to carry out the project, or must your project proposal compete with other project proposals that management is considering funding? If the customer is a governmental organization—a city, county, state, or federal (domestic or foreign) entity—there probably will be detailed specifications, formal quality standards, perhaps the necessity for surety bonds, and very often rigid and formalized inspection procedures. Who are the key personnel within the customer organization? Are they known to your organization? What history do you have with them or with the organization? Is your organization's reputation with the prospective customer favorable?

Your organization must have enough money to write the proposal and to sustain the postproposal selling and negotiating efforts. Therefore, you must know whether money to invest in this kind of activity is available. You must expect the project to earn more money than the proposal costs because you will not win every job on which you propose.

There are situations in which you will be the sole source recipient of an RFP or, for instance, when your boss tells you to carry out a project within your organization. There is thus no competition, but how you perform the job is still important. It may be better to decline an effort when it is offered sole source if you are convinced the performance you can provide will be at best marginal.

All competition has to be analyzed. Some relevant issues are the competition's technical and managerial competence, its ability to produce the requested project output, an estimate of its interest in the particular type of project, its need (or degree of "hunger"), and its prior relationship with the customer or organization.

Proposals are selling documents intended to help the customer discriminate one offering agent from another.

In summary, avoid projects that are inconsistent with your organization's long-term goals, inconsistent with your organization's current and near-term resources, unlikely to win the proposal competition, and unlikely to satisfy criteria established by the delivering organization for success (see Chapter 2).

THE PROPOSAL PROCESS

The proposal process is a series of nine steps:

1. Preproposal marketing and sales effort
2. Authorization
3. Selection of a dominant theme
4. Preparation of the statement of work
5. Development of a plan to satisfy the Triple Constraint (an effort for which a checklist may be helpful)
6. Adjustment to remove inconsistencies and inadequacies
7. Approval
8. Submission
9. Postsubmission follow-up, including presentations and contract negotiations

Preproposal Marketing and Sales Effort

Many newcomers to the world of contracted projects ignore the necessity of preproposal and postsubmission activities. The goal of the preproposal work is to learn about the customer's problem and bias. For example, requesters are often just trying to get three bids (or one bid and two no-bids) to justify awarding a contract to an already-selected supplier. Organizations waste many scarce resources writing proposals because they were not in the pre-RFP loop.

In preproposal marketing and sales efforts, you are trying to collect information on the customer's decision makers, the stated and unstated requirements, likely competitors, and likely partners and trying to develop your preliminary technical and management approaches. Good proposal organizations will have a substantial amount of homework done before the RFP ever crosses their transom.

Authorization

The marketing and selling process (preproposal effort) should start before the customer issues its RFP. Many organizations consider the proposal itself as a project with a Triple Constraint, in which case the performance objective is submission of a winning proposal and price, in accordance with the required submission schedule, for a cost acceptable in view of the probable (financial) return to your organization. If a project approach is used, this activity must be authorized, as we described in Chapter 2. A form such as that in Figure 4-5 can be used for this purpose.

In preparing a proposal, an organization is going to commit a certain amount of effort and money to it. This investment should be made only when it seems that the opportunity then available has a good chance of paying off, and only if it is consistent with the organization's goals.

Another point to consider at the time of proposal authorization is the individual who will manage it. Ideally, the proposal manager should be the intended project manager. Often, this ideal is not achieved. A proposal manager who knows that he or she will not be the project manager is likely to make commitments that are hard (or impossible) to meet, especially when the proposal manager's performance is judged on his or her ability to obtain (as opposed to manage) new business.

Theme Fixation

The proposal theme provides a focus to direct everyone contributing to the proposal, which increases the odds of producing a coherent, winning proposal. There are several reasons that theme fixation is important. In the first place, many customers are in fact organizations, and the "customer" is comprised of people who view the contemplated undertaking in slightly different ways. It is necessary to understand these subtle differences and either harmonize them or deduce who has the most influence.

In addition, the customer's statement of the problem in the RFP may be imperfect or incomplete. Early after receipt of the RFP, develop a list of questions, and submit them to the customer. Sometimes you can develop a productive dialogue with your prospective customer that will give you an opportunity to launch trial balloons representing your initial approach to the proposal. You will learn your customer's preference, which will permit you to adjust your thinking to produce a proposal that is more responsive to your customer's prejudices and predilections.

It is important for everybody in your organization to understand the chosen theme so that their contributions are consistent with it. This theme will be used throughout your written proposal. It might be technical sophistication and elegance, early delivery, the fact that the unit you propose to furnish is a proven item, or that you have a team ready to put to work on the job. If everyone in your organization understands what the theme is, your proposal should show your prospective customer that you are the best-qualified offering agent.

PROPOSAL AUTHORIZATION		NUMBER	REVISION
TITLE			

JOB

PERFORMANCE REQUIRED	
ESTIMATED STARTING DATE	ESTIMATED DURATION
ESTIMATED BID PRICE	ESTIMATED SUBCONTRACT TO OTHERS $ %
IS JOB FUNDED? WHAT IS FOLLOW-ON POTENTIAL?	
ESTIMATED NEED FOR CAPITAL AND FACILITY EXPENSE IF JOB IS OBTAINED	
CUSTOMER ORGANIZATION	
KEY CUSTOMER PERSONNEL	
CONTRACT FORM SPECIAL CONSIDERATIONS	SECURITY CLASS

COMPETITION

COMPETITORS
COMPETITORS' STRENGTHS
SIGNIFICANCE TO COMPETITORS IF THEY LOSE
OTHER COMPETITOR WORK FOR CUSTOMER
OUR UNIQUE ADVANTAGES

PROPOSAL

WHAT IS TO BE SUBMITTED?	
DUE DATE	PROPOSAL COST (DETAIL BELOW)
PROPOSAL MANAGER	OTHER KEY PROPOSAL PERSONNEL

EFFORT ACTIVITY	PROJECT DEPT. (HOURS)	SUPPORT GROUP A (HOURS)	SUPPORT GROUP B (HOURS)	SUPPORT GROUP C (HOURS)	NONLABOR (DOLLARS)
PREPROPOSAL					
BIDDERS' CONFERENCE					
PROPOSAL PREPARATION					
CUSTOMER PRESENTATION					
CONTRACT NEGOTIATION					
OTHER					
TOTAL HOURS					
TOTAL COST					

FUNDING NEEDED	JAN	FEB	MAR	APR	MAY	JUN	JUL	AUG	SEP	OCT	NOV	DEC
MONTHLY												
CUMULATIVE												

APPROVALS

GROUP A	DATE	MARKETING MANAGER	DATE	VICE-PRESIDENT—FIN.	DATE
GROUP B	DATE	DIVISION CONTROLLER	DATE	EXEC. VICE-PRESIDENT	DATE
GROUP C	DATE	DIVISION MANAGER	DATE	PRESIDENT	DATE
PROPOSAL MANAGER	DATE	VICE-PRESIDENT—OPERAT.	DATE	CHAIRPERSON OF BOARD	DATE

FIGURE 4-5. Typical proposal authorization form.

Finally, recognize that a proposal is a selling document. Your job in proposal writing is to win the work, not to plan or do the work. A common mistake is to include the entire work plan.

Requirements

Content The requirements [in some cases called a *statement of work* (SOW)] must describe the job to be done. It should designate any specifications that will be applied. It should identify measurable, tangible, and verifiable acceptance criteria so that there is no uncertainty whether the final item is in fact acceptable. It is also extremely helpful to know which dimension of the Triple Constraint is most important to the sponsor, which is second most important, and by implication, which is least important. Unfortunately, some sponsors have not thought about this, and sometimes sponsors will not provide such a priority ranking. Nevertheless, if you can obtain it, it will help you to make intelligent tradeoffs if these are required.

Figure 4-6 shows factors that influence requirements. From requirements, the team develops design specifications for products or services. Design specifications often require tradeoffs, as illustrated in Figure 4-7. Other common tradeoff examples include automobile carrying capacity can be enlarged substantially only at the cost of poorer fuel mileage and photographic film speed (light sensitivity) can be improved only at the expense of resolution. One conscientious person's perception may differ from another conscientious person's perception, so these tradeoffs must be resolved explicitly before team members begin to work at cross-purposes.

All requirements should be verifiable. Requirements specifications also have both quantitative and qualitative aspects, as shown in Table 4-2. Neither the proposal nor project teams have much trouble understanding how to judge the quantitative aspects, but they frequently get into trouble with the qualitative ones. Consider a product that must be

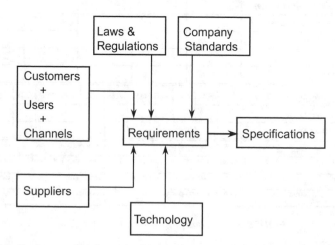

FIGURE 4-6. *Design specifications should satisfy requirements, which are affected by diverse factors.*

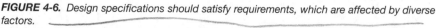

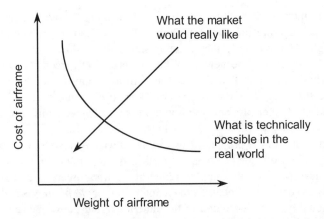

What the market would really like

What is technically possible in the real world

Cost of airframe

Weight of airframe

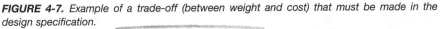

FIGURE 4-7. *Example of a trade-off (between weight and cost) that must be made in the design specification.*

"contemporary" or "attractive." Who decides, and on what basis? When forced to employ a qualitative specification, one must specify how satisfaction will be judged. It might be decided exclusively by the owner of the company, which is a common situation in a small or privately owned company. In a larger organization, many people may be involved. In either situation, the decision criteria must be understood, and judges of compliance must be kept informed.

As we wrote in Chapter 2, when it is not possible to be precise at the earliest phase of a job because the final product is not realistic or attainable, it is important to undertake a two-phase project effort. The first phase, perhaps extending to a customer review, is quoted completely, but the whole job is quoted only approximately, in a nonbinding fashion. The whole-job quote represents the proposer's best estimate of project requirements. The first-phase quotation, however, is firm (even if a cost-reimbursable contract form is used) and includes sufficient effort to construct acceptance criteria for the rest of the job.

Clarification You should review the requirements with the customer prior to further work on the proposal itself. Ambiguous words should be avoided, and it is desirable to be quantitative, using numbers and dimensions whenever possible.

TABLE 4-2 Illustrative Elements of Design Specifications

Quantitative	Qualitative
Size	Styling
Weight	Attractiveness
Speed	Aroma
Sensitivity	Taste
Resolution	Sound quality

A requirement such as "design and build an amplifier with greater gain than model L" or "explore the continental shelf for oil" can be ambiguous. The contractor might interpret the former as being satisfied by 10 percent greater gain. The customer might expect the new amplifier to have 300 percent greater gain. This leads to what could be called the "er trap." In too many situations, it is attractive to promise that a product will be faster, easier to use, smoother, lighter, roomier, sturdier, have a longer life, and so on without carefully determining what the "better" feature is truly worth to a user or buyer.

It is important for the project manger to avoid wasting resources on out-of-scope work—regardless of whether it is interesting or important. Figure 4-8 illustrates that the contract defines that narrow range of common interests that is to be dealt with in the project work. Recall from Chapter 3 that *gold plating* is adding features and functionality over that found in the functional and performance requirements. If you follow these two simple rules, you will avoid much aggravation:

Recognize out-of-scope work, and control it.

- Work to requirements.
- Recognize and manage change to scope.

Technical and Management Plan

Recall that the proposal is a selling document. In order to convince the customer that your organization is required, you need to convince him or her that you can plan, execute, and control the work. Your proposal will need to describe at some high level the technical approach and management plan.

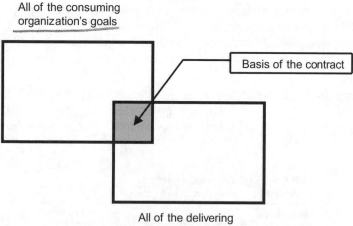

All of the consuming organization's goals

Basis of the contract

All of the delivering organization's goals

FIGURE 4-8. *The project manager must assure that other consuming organization and delivering organization goals do not distract the people working on the project from accomplishing the contract's requirements.*

Simulation   Simulation is used in many situations to get a better idea of what can and cannot be achieved. Servomechanism engineers, for instance, will simulate on paper, and perhaps further using a digital or analog computer, the performance of a servomechanism before attempting to build even the breadboard. In doing this simulation, they are investigating how the servomechanism might perform if it is built according to certain specifications.

Civil or structural engineers often will examine the deformation of a building or a bridge to determine if it will have adequate strength if built according to the design drawings. Similar simulations occur in many other fields, such as aerodynamics, thermodynamics, and optical design. A major purpose of these simulations is to identify any potential problem areas in the prospective system before building it. If a design won't work on paper, it won't work when it's built.

In the case of project planning, where the project plans are a simulation of how the project will be carried out, there are similar reasons for engaging in simulation. It is important to decide how to establish a price for the proposed work. A detailed simulation, that is, a plan, makes it more likely that the proposed or bid price will be sufficient. If the plan has been thoroughly prepared, it also will convince the customer that your organization understands the proposed job, which helps in negotiating your contract. If the project schedule is unrealistic on paper, the project won't be completed on time. Project management software can be used quite effectively for this schedule and budget simulation work.

The Triple Constraint   It is important to understand that the technical and management plans described in the proposal will become inputs to the detailed project plan. You use the work breakdown structure (WBS)✳ to describe your approach to the performance dimension. Many computer project management software packages have a WBS capability. You use network diagrams or, in some cases, a bar chart in which each activity corresponds to a WBS element to describe your approach to the time

The proposed work approach is not the same thing as the project plan.

dimension. You use a complete cost breakdown for each activity to describe your approach to the money dimension and defend your price.

These three planning elements are best prepared in the order presented. First, the WBS is used to describe the things that will be undertaken to satisfy the performance specification. After that is complete, it is possible to prepare a network diagram for each of these designated elements in the WBS. Initially, each of these items should be estimated in a "natural" time frame; then these activities' logical relationships to each other can be established. If, as usually happens, this produces an unacceptably long program, it is then important to decide which activities will be scheduled to be completed in periods shorter than the "natural" time. That is, some of the project activities must be carried out faster than is desirable. After agreeing to shorten the schedule, prepare cost estimates for each activity. Note that it is not desirable to prepare the cost estimates prior to determining the time to be allowed for a given activity. These issues will be treated in more detail in later chapters.

Checklists Checklists are designed to help ensure that nothing that will have to be dealt with during the course of the job has been forgotten or omitted from consideration in the proposal. A checklist should contain such items as shown in Appendix 3. These checklists are not exhaustive and may not contain the most significant or most important items. They are meant to suggest the kinds of items that might appear in a checklist.

The best way to develop a checklist is to create your own over a period of years. One way is simply to enter items on file cards whenever they occur to you during the course of project work. You may later sort out these cards alphabetically, by time phase, or by some other logical method. Having developed a checklist from your own experience, you will perform better on future projects because you will not forget items that are likely to be significant. Thus, you will consider their impact on a project during the proposal phase; they will not emerge as unexpected developments during performance of the project.

On a new proposal or project, think about it critically before consulting your checklist. If you turn to the checklist first, you tend to develop "blinders" to new crucial factors unique to the current job.

Adjustments

Adjustments are often required after a proposal has been partially prepared. Perhaps someone discovers that two departments contributing to the proposal have duplicated their efforts or have made differing assumptions about some significant item. Or perhaps someone discovers new information or corrects some oversight.

When an adjustment is required, all participants must join in deciding how to make it. Two benefits accrue from participants making the adjustment. First, the experts are considering the problem and presumably making the most sensible adjustment. Second, having contributed to the adjustment, other participants gain a sense of involvement in the decision and tend to perform the job better when the proposal has been converted into a project undertaking.

Approval

As with initiation of the proposal effort, the conclusion of the proposal requires managerial action within an organization. There typically will be a sign-off control sheet (Figure 4-9). Normally, each organization has a procedure that specifies the signature authority of given managerial levels, and such a procedure indicates which managers or officers must sign the control sheet signifying their approval for proposal submission. The sign-off control sheet must contain a brief description of the Triple Constraint contained in the proposal document being submitted. The sign-off control sheet is retained in the proposing organization's files and is not submitted to the customer.

It is important not to take the approval of senior managers for granted. Therefore, it is important to give these people timely briefings throughout the proposal preparation effort as to the scope of the proposal and the nature of the resources to be

PROPOSAL SUBMISSION APPROVAL	COMPANY PRIVATE	
PROPOSAL TITLE	NUMBER	
CUSTOMER	CONTRACT FORM	
SUMMARY STATEMENT OF WORK		
SCHEDULE FOR JOB		

COST	FEE	TOTAL BID PRICE

DOCUMENTS, REPORTS, MODELS, ETC., SUBMITTED
SUMMARY OF OUR RISKS
KEY PEOPLE PROMISED
FINANCIAL COMMIITMENTS REQUIRED
WARRANTY
ACCEPTANCE CRITERIA
REMARKS

MARKETING MANAGER	DATE	VICE-PRESIDENT—FIN.	DATE
DIVISION CONTROLLER	DATE	EXEC. VICE-PRESIDENT	DATE
DIVISION MANAGER	DATE	PRESIDENT	DATE
VICE-PRESIDENT—OPERAT.	DATE	CHAIRPERSON	DATE

FIGURE 4-9. *Typical proposal submission approval form.*

committed to the resulting project. Although the proposal authorization document (Figure 4-5) constitutes one such involvement of senior management, it alone will not suffice. The number and frequency of such briefings during proposal preparation depend on the organization, its rules and procedures, and the proposal manager's good judgment.

In the case of proposals to government customers, there is usually a requirement for an officer to sign a certificate of current pricing or similar form. This is intended to assure the government that the proposed price is calculated in conformance with current laws and procurement regulations. Because the signer of this certificate is subject to substantial personal penalties if the certificate is inaccurate, this officer will require a careful review of your technical and cost proposal.

Submission

The time comes to submit the proposal to the designated recipient, who may require that it bear a postmark by a certain date or be received at a given office by a specified date stamped by a particular time. Such standards are overriding and must be complied with.

As a practical matter, proposals normally should be submitted in three sections:

1. Executive summary
2. Main proposal (dealing with both technical and management issues)
3. Appendices

Each of these sections actually might be a separate volume, and in some cases, the technical and management portions, as well as each appendix, might be separate volumes. The exact content of each part may be stipulated by the RFP. If you generate a proposal on your own initiative, you may have more freedom to include what you want in locations that you judge will be best for the intended recipient.

Postsubmission

Mailing or delivering the proposal is not the end of a winning proposal effort. There is postsubmission work. At the very least, the winning organization must negotiate a contract with the customer. Sometimes several proposing organizations are deemed qualified, and negotiations are carried out with two or more of them prior to selecting the winning contractor.

In many proposal situations, the negotiation phase is preceded by a presentation to the customer. Such a presentation may be elaborate, requiring special graphics and models, and may entail extensive time and effort. The proposal contains a statement of work, which is the basic project definition. To prepare the proposal, it is necessary to do some, but not all, of the project planning. Much of this planning is commonly included in the proposal.

To summarize, the following are steps in the proposal process:

- Establish the organization's business strategy.
- Understand the organization's resources.
- Get the authorization (bid/no bid).
- Make the preproposal effort.
- Receive the RFP.
- Attend the bidder's conference.
- Establish the theme.
- Prepare the statement of work.
- Plan the job.
- Adjust the proposal.
- Approve the proposal.
- Submit the proposal.
- Present the proposal to the customer.
- Negotiate the contract.

TYPICAL PROBLEMS

There are both practical and people problems with proposals for projects. The first practical problem encountered in proposing winning projects is attempting to do virtually the entire job during the proposal. That is, in trying to prepare a solid proposal, you spend too much time working through the plan for the project. Sometimes this includes doing preliminary engineering, modeling, or program coding. You can overcome this by recognizing that risk must be balanced and planning only enough to reduce the project's uncertainty to an acceptable level.

A related problem is inadequate project planning in the proposal. The solution here is to keep planning until it becomes too time-consuming. This is obviously a judgment issue for which personal experience must provide guidance. Another problem is the last-minute rush to complete the proposal in time to submit it. The solution to this problem is to have a proposal preparation schedule and adhere to it. Another frequent problem is a poorly crafted specification, regardless of whether the proposal is initiated in response to an RFP or self-initiated. The solution is to be explicit about both the quantitative elements and the means by which the qualitative elements will be judged.

The principal problem with people is getting the workers who will subsequently perform the work on the project to contribute to the proposal. The ideal at which to aim—seldom completely achieved in practice—is to have all the people who will be key project participants describe, plan, and estimate the work they will later do.

INTERNATIONAL PROJECTS

International projects typically have added levels of complexity that you should be aware of.

International projects are not fundamentally different from domestic projects. In practice, however, the travel lure of a remote and apparently salubrious destination seems to affect human judgment. Furthermore, international projects expose a contracting organization to special problems related to language, currency, and unfamiliar business practices.

Remoteness

The other party in an international project typically is located in some remote area of the world. The far-off hills often look green. When two project opportunities are available to a company or organization at the same time, the one that originates in Zurich (or some other location you want to visit) will receive far more attention than the one that originates in Cleveland (or someplace to which you are averse), and this is independent of the intrinsic merits of the project opportunity. This is the *travel lure,* and it is a problem you must recognize and identify for what it is.

There are occasions, however, when you should undertake a project involving a foreign customer or partner. When this is the case, it is desirable to get more information about the foreign organization or the business in which it is engaged. If your organization has an office in that country, this is the best source of information. The foreign organization's country may have an information office in the United States, which may be able to provide useful information. The U.S. embassy or trade centers in the foreign country may have information, as may the U.S. Department of Commerce or the World Bank.

Having gathered all the information you can without leaving the United States, you then may have to succumb to the travel lure and go to the foreign country to discuss the project or negotiate. It is generally prudent to go several days before the initial meeting to compensate for jet lag.

Business Practices

The business practices governing project performance as well as negotiations and discussions leading to the project typically will be those dictated by the customer's country. For instance, Japanese tend to discuss a project at great length to try to achieve consensus. When this has been done, a handshake or verbal agreement is binding.

Staffing requirements frequently will require that some of the work be reserved for the customer's nationals. Sometimes nationals of a third country are stipulated or prohibited for a variety of reasons. Permits and "red tape" enter into business practices, and these often will absorb far more time than expected. In some cases, export or import controls can impose severe restrictions, so it is always important to understand these issues early.

Laws and Regulations

There are numerous laws and regulations, unfamiliar to the vast majority, that apply to international business. Limitations on the export of technology from the United States make headlines periodically when someone or some company is charged with a violation. Such violations typically involve shipment (or indirect reshipment) of some restricted high-technology product to an unfriendly nation.

In other situations, even more arcane restrictions, such as the Export Administration Act, may snag the unwary project manager. In one case, a U.S. congressman allegedly broke a law prohibiting U.S. citizens or companies from aiding the Arab boycott of businesses tied to Israel merely by writing a letter to someone in Kuwait. Thus, if you are contemplating an international project, be sure to have competent legal guidance on what you can and cannot do.

Language

Fortunately for people whose first language is English, most contracts are written in English. In some cases, however, a customer's language will prevail, and the contract will be written in it. If you are using a translation, it is prudent to have several prepared because the original host-language document is the controlling document, and interpreters are likely to interpret it differently. If you find wording differences, you can explore their significance with a language expert.

Price

As with language, the customer's currency may be the stipulated medium of financial settlement. Because currency rates fluctuate, it may be desirable to insure your company against these fluctuations. This is done with a hedge contract. Alternatively, you may use letters of credit to satisfy the future payment obligations. In any event, your price must allow for the extra expense of doing business at a great distance.

HIGHLIGHTS

- Contracts are agreements that can be informal (e.g., verbal) or formal (i.e., written).
- Government contracts include numerous special regulations.
- A proposal should be made only if a reasonable contract can be negotiated. There are several contractual forms, including fixed price, firm fixed price, cost plus fixed fee, cost plus incentive fee, and time and material.
- Both parties must be prepared to make concessions during negotiations.
- Organizations must apply their portfolio management strategies to filter out opportunities that have a low probability of success.

- A proposal is a sales offer showing that the offerer is willing to enter into a contract.
- Four issues involved in the decision to bid on a proposal opportunity are the nature of the requirement, the value of the project, the organization's response ability, and its ability to win.
- A checklist may help you to avoid overlooking required work.
- International projects introduce special problems, such as unfamiliar business practices, laws and regulations, language, currency, and distance.

Part 2

Planning a Project

5

Planning the Project

Planning is a process that people use to share their different views of the project goals and methods. Through planning, the project team develops a model of the project, which they often formalize as "The Plan."

The project integration management knowledge area (introduced in Chapter 1) includes the activities that are concerned with planning and executing the work of the project. Planning is an integrative process, and the plan is one artifact of the planning process. Successful project managers will stress the planning process because a good planning process will lead to an organized, valid plan. The following four points are particularly significant:

1. Determining where you are now (or will be, if you expect the project's "notice to proceed" at a later date)
2. Describing your goal (We often ask people to work on this question: What does "done" look like for your project?)
3. Selecting the approach that optimally gets the project from its current situation to the required future state
4. Determining your tolerance for variations from "plan" and techniques to control that variation

We illustrate these four factors in Figure 5-1, which uses a common project management graphic, the *S-curve*. First, notice that the two axes of the graphic show time elapsed and cumulative work complete. At the start of the project, the project has not performed any of the planned work. Second, examine the thick line designated as "Plan (Baseline)." The answers to "Where you are now?" and "What does done looks like?" are the two end points of the plan. Third, the shape of the S-Curve tells you something about the approach for accomplishing work. For example, you could "front load" the project and try to get a lot done early. Alter-

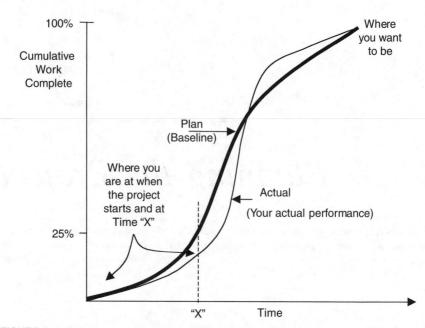

FIGURE 5-1. *Planning establishes the route to get from where you are at to where you want to be.*

natively, you could "back load" work and plan on finishing strongly. Fourth, note the point at time "*X*" while the project is underway. In this case, it appears that you have actually completed about 20 percent of the work, whereas you had planned to accomplish 25 percent of the work at time "*X*." In this example, you have a variance that you need to manage to nudge your performance back toward the baseline. Fifth, note that the "Actual" performance is better than plan at a later period (when the "Actual" line is above the "Plan" line), and you may choose to correct that positive variance. This S-Curve graphic summarizes the essence of the practice of "planning your work" and "working your plan." To reinforce a point made previously, "Projects don't fail at the end; they fail at the beginning," this chapter is of particular importance to successful project management.

INTEGRATED PROJECT PLANNING

Integrated project planning is a process that involves all the knowledge areas of project management (see Chapter 1 for a review) and helps to ensure that the project has a balanced, systematic approach to achieving the results. The following paragraphs describe several important concepts of integrated project planning, and Figure 5-2 illustrates the general flow.

Many project managers are asked—or feel compelled—to take on responsibility for a project that another team planned. Here are some suggestions:

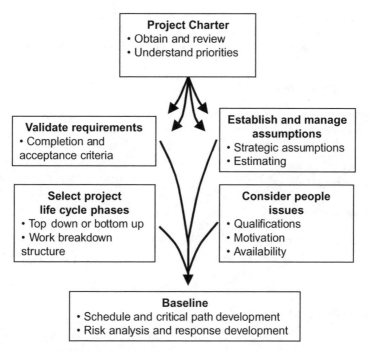

FIGURE 5-2. *Simplified flow of integrated project planning.*

- Evaluate the team members' qualifications and motivations. Do you have the right people involved?
- Take some time to study the project's requirements, as well as the product and the work scope. Meet with the previous project manager and project team. Even if the project manager is different, typically, some people will be carried over from the previous project team.
- Evaluate assumptions.
- Develop options, and discuss them with stakeholders.
- Conduct a risk analysis.
- Renegotiate as appropriate.

Review Project Authorization Documents

Because the scale and complexity of an endeavor are often unclear initially, you should ensure that you are undertaking a project and not a task. One easy way to do this is to determine if there is a project charter, as we described in Chapter 2. The project charter should alert you to the organization's project portfolio management strategies and goals. Information in the charter can provide important context for your planning discussions. If a project charter is not present, you have an issue that you need to manage, and we recommend reviewing some of the points in Chapter 2 and the issues management process as described in Chapter 11. You also might want to review the benefit-to-cost or discounted-cash-flow calculations that are part of the projects business case (see Chapter 9).

Validate Requirements

Continue to capture and understand requirements while planning and executing the project. In looking back over projects, most experienced project managers will tell you that many of their greatest frustrations originated with initially poor and then changing requirements. You need to recognize the imperative to capture requirements as an ongoing issue during planning and execution of the project to develop specific, measurable acceptance criteria. In their haste to "get moving" with a project, people tend to make assumptions rather than ask questions and often work without a requirements specification. Good project managers will continue to allocate work to managing the product and project requirements and will anticipate change.

Consider People Issues

Recruit the best people for your project. Most good project managers recognize the importance of getting the right people involved in the planning process and getting them involved early. In every organization, there is a small cadre of people who are exceptionally productive individual contributors. You want to recruit those people to your project and avoid accepting "whoever is available" if they can't do the job that you need to have done. Further, some projects involve considerable ambiguity, and you want people who tolerate it and take some initiative. You should actively recruit the best-quality people you can for your project, and you should read Chapters 13 and 14 for advice on team building.

As an aside, since all projects are competing for the best people, management's resource-assignment decisions reveal its underlying portfolio priorities. Since the best projects should get the best resources, you simply need to identify the projects that get the best people to discover priorities. For highly technical work, the optimal workload is approximately two or three projects per individual. Higher workloads dilute the time that the individual can spend on any project.

Many project managers try to "go it alone" and do all the planning themselves because they have been told, "People don't like to plan." They believe that this approach is efficient and will reduce the frustration for others. However, this is a poor approach because people are not likely to support a plan for which they provided little input. Consider this comment from a team member who was not involved in the planning process and was a reluctant, unenthusiastic, and frustrated contributor to a project:

> As a team member responsible for assembling instrumentation and data acquisition equipment for project use, I face a recurring problem with some project managers.

There seems to be an almost universal belief among members of that group that instrumentation issues are of little consequence and that any problems that may occur will be easy to resolve. When I am given a project assignment, I often find that little or no thought has been given to instrumentation availability, cost, or possible lead time.

The people who are the "planners" should be the "doers" and the "doers" should be involved in planning. Good project managers involve the doers with the planning process because

- They probably know more about it than anyone else.
- It is *their* task, not *yours*.

You could consider the preceding prescription to involve the doers with the planning process as something of a golden rule, and we will refer to this prescription as the *golden rule* throughout the remainder of this book. While it is not exactly the same as the moral principle expressed in the Golden Rule of "Treat others as you would want to be treated," there is a similar sentiment.

A corollary to the golden rule: No job is too hard for the person who does not have to do it.

Since a project is an organization of people, you should recognize that each person has rational but different mental models of the project's goals and the approach to achieving them. They can stubbornly hold onto their ideas and cause frustration. It is better if you regard planning as a social process: a conversation among various stakeholders about the future of the project. Thus, planning is a process of researching pertinent information, building and testing assumptions, and integrating them into a communications strategy. As we described in Chapters 1 and 2, an important part of successful project management is setting and managing expectations.

Select the Project Life Cycle

By definition, projects are progressively elaborated. One early planning activity is that the project manager and team should select the phases that will constitute that project's life cycle. For more straightforward projects, the project team might use a set of phases that would include further requirements definition, design, construction, and closeout. In these projects, the phases provide a structure for determining completion criteria.

The project life cycle provides an integrating framework for the project.

As another example, someone may request a preliminary estimate for budgeting and building a business case. The need for the budget estimate could precede the definition of detailed project requirements by one or more years. In this case, the project team would recognize that assumptions generated early in the project life cycle might be invalid. Still, some managers will take the early assumptions as the basis for their budget and establish that as a funding constraint for the project. Too often, the budget is insufficient. If this happens to you, identify it as an issue, and escalate to the project sponsor. (See Chapter 11 for more description.)

In many organizational environments, there are many unknowns, lots of detail and complexity, and everything is changing fast. To respond, some organizations use the rolling-wave technique in their project life cycle. The rolling-wave project plan is a sliding window of planning and execution in which the project uses an iterative process characterized as "plan a little, do a little." In this rolling-wave technique, each window's duration is typically about three months. As the team moves into each window, it decomposes activities and updates the planning baselines. In rolling wave, the project team does not try to develop a detailed estimate of activities that fall outside the current rolling-wave window. While rolling wave requires some sophistication by the project and its sponsors, it provides an agile, fast, and flexible approach to delivering project value. Conversely, it also contains an inherent risk of project elaboration and potentially increased scope that must be guarded against.

The work breakdown structure (WBS) is an important activity that we describe in Chapter 6. The WBS is derived from the product requirements specifications and the product design. The WBS identifies every work package that must be completed to satisfy the requirements.

Project managers and teams commonly uses the phrases *top-down planning* and *bottom-up planning*. Top-down planning refers to the process of taking a goal— such as a budget or due date—and decomposing the elements of the goal into smaller chunks. For example, say that the project's goal is to build an addition to a warehouse, and management has budgeted $100,000 for the project. A top-down approach to planning would be to allocate parts of that budget to excavation and concrete work ($20,000), to the structure and roof ($40,000), to electrical ($15,000), to plumbing ($10,000), and to miscellaneous ($15,000). The project team would further divide each of these categories down to controllable work packages that they would budget to the individual performing groups. They would follow a similar process for the schedule: taking the time available from the present to the due date and fitting the required work into the time allowed.

The bottom-up planning approach, by contrast, starts with an analysis of the product requirements specifications, which may be found in a document of the same name or in related documents such as the SOW. In bottom-up planning, the team would identify all the required work and develop estimates of time and duration for each of the individual work packages. The team then would assemble the individual estimates and "roll them up" to the project level to determine the project's estimated cost and duration.

It should not surprise you to find out that good project managers will use both top-down and bottom-up approaches to planning to ensure that they have a complete and valid understanding of the project.

A project plan is not a "to do" list!

One important purpose of project planning is to establish the project control philosophy. You need to consider granularity of your control in your project budget or schedule. We have seen many instances of project plans resembling a "to do" list. A "to do" list may be appropriate for an individual managing a task, but it is too much detail for a project and sets the project up for micromanagement (defined as telling someone how to do their job rather than what their requirements are).

Establish and Manage Assumptions

Since we do not know the future with certainty, we need to make educated guesses. This educated guess is called an *assumption.* An assumption is a planning factor used to build the model of a project's performance and typically includes factors such as duration, cost, and risk. We cannot conduct project planning without making assumptions. Note that people can and will change assumptions:

- It is important to validate assumptions to ensure that they are representative,
- Documentation helps the project team to recognize early and agree on the validity of assumptions.

Assumptions are necessary for project planning, and they should be documented and validated.

It is helpful to recognize two categories of assumptions: *Strategic assumptions* are assumptions that affect the "go" or "kill" of the project. These assumptions are often found in the project's business case and charter documents. For example, most new product development projects include estimates of market size, market growth rates, pricing, and competition. The assumptions found in some business cases are frequently nothing more than wishful thinking. Better organizations try to evaluate the validity of their assumptions and kill their projects early. We even know of

Strategic assumptions are the basis for the project business case.

firms that have a metric called *wasted-development spending* that measures the amount of money wasted by projects that continued beyond the point at which they made economic sense. In these firms, project managers are part of the decision-making team and are prepared to recommend terminating projects based on assumptions that are no longer valid.

Estimating assumptions are the second type, and they allow the project team to "size" the project in terms of cost, duration, product quality, and risk. Examples of estimating assumptions include the following: availability of people to work on your project, time to complete a task, expenses associated with a task, and learning-curve effects.

The schedule depends on the resources that will be committed, so you need to recognize how your planning assumptions interact. The resources available for a project depend on priorities and the resources dedicated to other work. Consequently, portfolio management strategies for allocation of resources might delay a given project.

The project team will be performing resource planning and cost estimating for each task. The resource plan includes both human and physical resources. For instance, a senior person with relevant experience should be able to carry out a specific task more quickly than an inexperienced junior person. Similarly, if certain resources are overcommitted, the schedule will be unrealistic, and you will have to adjust your planning assumptions.

Consider the planning matrix tool illustrated in Figure 5-3. It lists project activities along one side of a piece of paper and designates involved personnel along the perpendicular side. Where these rows and columns intersect at a checkmark, the designated personnel are involved in the designated activity. For instance, a carpenter works on wood framing, not the plumbing or electrical wiring, in a construction project. Similarly, an expert on electronic circuit design works on the electronic circuit design task, not on the optical design task, for an electro-optical system. Project participants need to understand what management expects of them and what others are doing.

Baseline

For the baseline, the project team develops an understanding of the predecessor-successor relationships through a network diagram, as we describe in Chapter 7, that adds its cost and resource estimates, considers risk, and makes its plan final.

The project team should engage in project risk management and issues management early in the project. Chapter 11 covers this and more, and it describes a process of risk identification, risk analysis, and risk-response planning.

USING COMPUTER SOFTWARE DURING PROJECT PLANNING

Critical thinking is essential to project planning. Do not make the common mistake of assuming that project planning is simply a matter of filling out a software tem-

Task or Activity \ Function	Marketing	Engineering	Quality Assurance	
Market Research	X	X		
Product Specification	X	X	X	
Preliminary Design		X		

FIGURE 5-3. Partial example of a planning matrix, showing involvement of groups in project activities.

plate. Nor should you believe that the software has artificial intelligence that will do your planning for you.

Based on experience, we find it better to use a pencil-and-paper approach to project planning before you start entering data in the software. An alternative to using a computer is to have the key project team members use a roll of wide paper affixed to a wall, attaching "sticky notes" to show predecessor-successor relationships. The reason is that it helps the team construct a project model that everyone can understand rather than simply focusing on the artifact of a Gantt chart or some other kind of document. Once the basic project model is developed, start by entering the task names, durations, and resources into the software. You should think through the process very carefully to ensure that you are using a consistent calendar, time intervals, resources names, cost rates, and so forth.

The computer gives you the advantage of running numerous iterations and making updates during the planning process. Aim your first attempt at getting all the tasks and their interdependency listed without regard to the duration of each task. A second iteration modifies the first by including expected times (or calendar dates) for each task. A third iteration adjusts the second to reflect realistic resource availability. You use subsequent iterations to optimize the project model.

Remember that your goal is to drive the planning to a baseline "freeze" that becomes the budget during execution and control. You should save your computer file under a baseline so that you can monitor deviations from the baseline during the executing portion of the project. Recognize that a common mistake is using the computer is to make frequent minor revisions after the baseline decision.

Finally, if you do make changes to the baseline, be sure that you are practicing version control in your documentation strategy. Make sure that all planning documents and revisions have a revision serial number and date. It is important to know who has copies of documents so that if you revise plans, you can provide revisions to all the people who have copies of previous plans.

Make sure that all copies of your project plan are current so that all your project team is working on the same plan.

"THE PLAN"

Project plans may vary from simple one-page statements to records with extensive, intricate levels of detail. There is an appropriate level for each project undertaking. In many large projects, there is a book called "The Plan" or "The Project Plan." You can find it in a corporate information system or simply as a series of notebooks.

"The Plan" is the document that formalizes the project model and agreements. As an important subprocess of project management, communications is concerned with who receives information when and in what form. Documentation is a part of project communications management. An effective project plan accomplishes the following purposes:

1. It identifies the important work required to complete the project successfully.
2. It captures and documents assumptions and agreements.

3. It is credible to both the planners and management.

4. It facilitates effective communication by having an appropriate level of detail for the audience.

Your project plans matter. Even if you can perform your project in your office, other people in the organization (e.g., your boss) will want to know where your project is headed, what you are doing, and how long you will be doing it. Thus, project plans constitute an important communication and coordination document and may motivate people to perform better. The project manager should disseminate "The Plan" to all involved personnel. In very large projects, dissemination may require a chart room in which the project team post charts on the walls displaying the plans for and status of various activities, including financial progress and resource allocation. Of course, there are many good options available to use Internet and intranet capabilities to practice what people commonly call *Web-enabled project management.*

Topics Covered

Typically, the document will include the following topics:

1. Project summary
2. Project requirements
3. Milestones
4. Work breakdown structure
5. Network diagram of the activities with schedule dates
6. Budget for all activities
7. Project management and organization charts
8. Interface definitions, including facility support
9. Logistic support
10. Acceptance plan
11. Standards for property control and security
12. Customer organization contact points, if relevant
13. Nature of project reviews

APPLYING PROJECT PLANS DURING EXECUTION

A common adage for project managers is, "Plan your work, and work your plan."

Project planning and project plans have functional value to the project team. They provide a common reference point that is a basis for monitoring, control, and corrective action. In Chapter 6, we will describe the concept of the *work package,* which is a unit of work with describable requirements, resources, risks, duration, and so on. *Work authorization* is the term for authorizing the start of work on a work package.

Project plans require activation. You should obtain whatever higher-level approvals are required, including those of the customer. We want to point you to two important points for activation. The first is the baseline, which we have already discussed in detail. The presence of a baseline shows that the delivering organization is prepared to start executing the project. The second point is the customer's *notice to proceed,* which gives the project team permission to start the work.

As we illustrated in Figure 5-1, plans are also the basis of your project monitoring and control activities during project execution. It is a truism that projects do not go in accordance with plan. What you do not know when you start is where and how your project will deviate. Good **You must be alert for deviations from plan.** project managers will monitor progress and stay alert for early warning signals of deviations from the baseline. If necessary, they take corrective action.

Figure 5-4 illustrates how project planning documents are used for supporting project monitoring and control activities.

PROJECT PLANNING IS AN INVESTMENT, NOT AN EXPENSE

Sometimes people perceive that there is no value provided by planning or plans. They resent demands from a powerful person such as customer or a boss. However, most people agree that planning is a useful discipline despite its imposition on their own personal time. Here are a few important points to recognize:

- Planning is an investment that reaps benefits. Abraham Lincoln reputedly said, "If I had eight hours to chop down a tree, I'd spend six hours sharpening my ax."
- One benefit is in avoiding waste. Most people recognize the truth behind this statement: "We don't have enough time to plan now, but we'll have lots of time to fix it up later."
- One of the major frustrations of project participants is poor communications, so documents help to mitigate this complaint. Documentation has value in ensuring that communication between parties is consistent and understandable.
- Planning's major benefit is in avoiding failure consequences. Project planning does not guarantee project success or even cause it. (Most people have experienced a success in the absence of planning. The important nuance to recognize is that every time that we hear of an unsuccessful project, we find that people complain about the lack of planning.)
- As project complexity increases, the value that planning adds also increases.
- Experience shows that projects and organizations are more likely to underplan than to overplan.

HIGHLIGHTS

The process of planning is one of the most important elements of project management, and there is a large tool chest that we will expose you to in the following

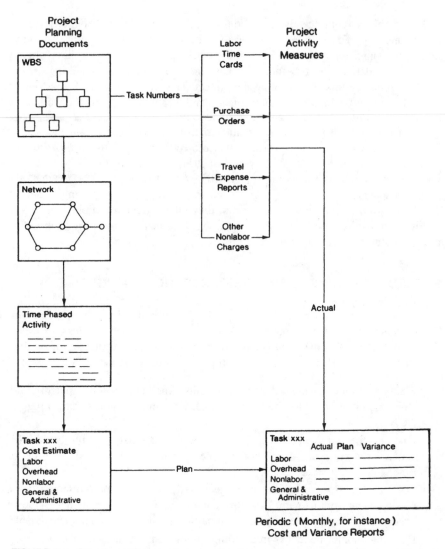

FIGURE 5-4. *Overview of how project planning documents are used in project and control.*

chapters. It is also important to recognize that the plans generated by the planning process increase the probability that people will have a common understanding of the project. We would be remiss if we did not remind you that leading and influencing people are essential project management activities.

- Planning is a social process of learning and discovery.
- The "plan" is an artifact of the planning process.
- Documentation has value—it forces clear thinking and makes communication more understandable.

- Consider your desired "level of control" as you establish the granularity of the planning.
- When top-down planning is used for budgeting, you can decompose work and attach budgets for time and cost. Plans delegate portions of the Triple Constraint to the lowest reporting level.
- Everyone involved must receive every plan revision.

6

The Work Breakdown Structure

Since project management is a process of managing work to create a project result, good project managers include identification and organization of the work activities early in the planning process. This chapter describes two essential integrating concepts for managing the project's work: the work package and the work breakdown structure (WBS). Work packages partition the project work into controllable bundles. The WBS is the project management tool for listing and organizing all the project's work. Both are important inputs to developing time and cost estimates.

In the preceding chapter, we discussed the processes of project planning, including the phases that constitute a project life cycle, the need to control activities, and documenting assumptions. In this chapter, we will describe the essential planning practice of creating the work breakdown structure (WBS). In particular, we will describe the work package as the fundamental unit of performance and show you how to create a good WBS. Finally, we will tie in this chapter's discussion to a central theme of *Successful Project Management:* the role of integration in the project's performance.

THE WORK BREAKDOWN STRUCTURE

The best way to explain what a WBS is to provides some examples. Figure 6-1 is an example of a WBS for a photovoltaic solar power system. In this example, there are two levels of decomposition below the top level labeled "Photovoltaic Solar Power System."

For another example, examine Figure 6-2, which illustrates a partial WBS for a bicycle product launch. Note the cascading levels of detail and how the numbering makes the hierarchy explicit.

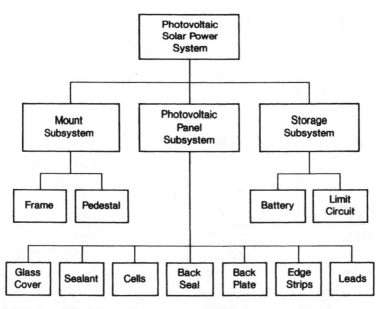

FIGURE 6-1. Example work breakdown structure (WBS) for a photovoltaic solar power system.

Figure 6-3 is a WBS in an outline format and is the way several software packages accept WBS information. Note that the first and last entries ("START" and "DONE") are milestone events; thus, they require no time or resources. Every project should include start and finish milestones as shown here but also may include others (such as customer approvals at certain points). Note the four summary groupings and the ten tasks. Technically speaking, milestones are not work packages because they have no duration, use no resources, and are not work.

The notions of hierarchy and decomposition are common to all three of the examples. In each case, there is a "project" level and two lower levels. We call the first level below the project level *level 1*, which aligns with the time phases or major costing categories. In these three examples, *level 2*, or the lower-most level, contains the work packages, a concept we will describe in more detail in the next section.

Use a WBS to organize all the identified work.

THE WORK PACKAGE AND THE WBS DICTIONARY

By definition, the *work package* is the entity at the lowest level of each branch of the WBS. The work package includes the schedule activities and schedule milestones required to complete the work package deliverable or project work component. Take another look at Figures 6-1 through 6-3, and respectively, you will see listed 11 work packages, 9 work packages, and 10 work packages.

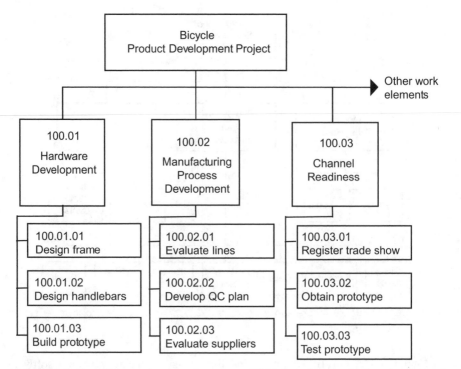

FIGURE 6-2. Example work breakdown structure (WBS) for a bicycle product development project.

Work packages are the fundamental units of a project.

Work packages are components of the project by which the project manager and participants plan, execute, and control the work of the project. Work packages are also known as *tasks* or *activities* when they are described in the context of project time management.

The WBS dictionary contains detail about each task.

The *WBS dictionary* is a document that provides descriptive information about each work package, including a brief description of the requirements, deliverable(s), work package identification number, and targeted start and end dates. It can include other information, such as task leader name and responsibilities, estimates of cost and resources, quality requirements, and technical and contract information.

TOP-DOWN PLANNING APPROACH FOR DEVELOPING THE WBS

Recall from Chapter 5 our introduction to top-down and bottom-up planning. The top-down approach to developing a WBS tends to focus on results and deliverables, whereas the bottom-up approach helps the team to develop a detailed understanding of the work activities.

ID	Task Name
1	START
2	Materials Tasks
3	A - Select Materials
4	B - Obtain Materials
5	Equipment Tasks
6	C - Build
7	D - Debug
8	E - Conduct Experiment
9	F - Document Design
10	Theory Tasks
11	G - Review Literature
12	H - Make Theoretical Study
13	J - Write Theoretical Report
14	Report Task
15	K - Write Final Report
16	DONE

FIGURE 6-3. *A WBS in outline format.*

The top-down approach starts with a "big picture" understanding of the goals and objectives of the project. In the top-down approach, the project team builds the WBS by partitioning the work (decomposing it) into smaller work packages. When reviewing a completed WBS, you should be able to determine that each descending level represents an increasingly detailed description of the project elements. Thus, the WBS aids in aligning work to the top-level vision. It also assists the project team in verifying that requirements have been met and thus validates the work scope of the project.

ORGANIZING THE WBS FOR COMPLETENESS AND CONTROL

Top-down approaches are especially useful for organizing top levels of the WBS as a prelude to bottom-up planning and estimating. When the top levels of the WBS are done well, the project team will develop confidence that it has identified all the work and can manage and control the work in accordance with its control philosophy.

You need to be able to identify the work before you develop schedules and cost estimates for it.

A complete WBS ensures that the project team has identified all the work and thus has the necessary correct inputs to develop the schedule and cost estimates. When organized according to its control and execution needs, the WBS ensures that all the required project activities are logically identified and related. Returning to Figures 6-1 through 6-3, note the important implication that the WBS displays each work package for the project. In other words, you only perform the work that is on the WBS, and if there is a change to the scope of work, you track it through the WBS.

In applying a top down approach, the project manager and team will choose one of the following high-level structures:

- *Product or system.* Developers of large hardware systems often use this approach to defining level 1 of the WBS. Figure 6-1 is an example.
- *Functional department.* This approach is used often for internal projects where the focus is on what work is required from a given department. A variant of this style is to organize work geographically. Figure 6-2 is an example of a WBS organized by functions.
- *Time-phased.* The project team elects to organize the project work according to the phases of its selected project life cycle. Many organizations mandate a standard project life cycle and hence a standard WBS. They believe that this standardization makes the process of project planning easier and more consistent. Figure 6-3 shows a general sequencing of work from beginning to end within the materials, equipment, and theory tasks groupings.

Variants and hybrids of these types of WBS organizations are common. Further, the level of decomposition varies with the complexity and control needs of the project. As a general rule, you should decompose the project to the third or fourth level, but it could be appropriate to show five or ten or even more levels.

Do not blindly copy another project's WBS for a new project without considering the differences between the projects. If the differences are not resolved, you may waste everybody's time—the people who prepare the WBS and the people who use it. Prepare the project's WBS thoughtfully, not by rote, to increase the odds of project success.

BOTTOM-UP PLANNING APPROACH FOR DEVELOPING THE WBS

The bottom-up approach to developing the WBS is based on the premise that numerous time- and resource-consuming activities are needed to yield a given result or deliverable. It is widely believed that the work performers have the best understanding the work and are in the best position to identify the work and estimate the resources and cost needed to accomplish it.

Note that Figure 6-1 includes the hardware for a photovoltaic solar power system, but the WBS does not list a shipping container, the installation and operation instructions, a warranty document, or a user training manual, some of which might be a part of such an engineering project. A bottom-up planning approach probably would help to flesh out the necessary detail for the work packages or the WBS dictionary.

Here is one common approach to bottom-up development of the WBS. Assemble the project team and describe the project's requirements, goals, and assumptions. Next, pass out packets of sticky notes, and instruct people to first write the names of the work packages on individual sticky notes. Actually, you can use different colors of sticky notes, where each color is work done by a particular organization (e.g., division, section, department, subcontractor, and so on). Typically, the project manager has selected the level 1 structure of the WBS, so next, the team arranges the work packages into the WBS hierarchy. This is a good time to remind the project team to remember tasks that do not result in deliverables, such as reports, reviews, coordination activities, analyses, and tradeoff studies. Displaying them on a WBS is a good way to highlight that they are necessary and that resources must be devoted to them. Developing a WBS in this fashion is easy and offers great benefits in creating understanding about project scope, roles and responsibilities, durations, estimated resources and costs, risks, dependencies, and critical decisions. In addition to getting a good WBS, people gain a better understanding of the big picture of the project and are supportive of the project.

Most people need to practice decomposing work into smaller chunks. Participants have to train themselves to look for several things: deliverables, or logical chunks of work, or dependencies between project modules. As they develop this skill, the quality of their WBS is noticeably improved.

Develop skill in partitioning and organizing the work.

Work Focus, Not Product Focus

When using a bottom-up approach, we recommend that the team name work packages with the "verb + noun" convention. For example, "Design User Interface" is much more descriptive than "Interface" (which could be confused with "Test the Interface," "Rework the Interface," or similar tasks). Each verb adds a new work package. Some people argue that the verb-noun format is superior because it emphasizes that the project is tracking work, thus avoiding the common pitfall of getting "hung up" in technical problem solving. The verb is the "action," and the noun is the "object of the action." Others argue that the verb noun format is not necessary. Compare Figure 6-1 and 6-2, and note that Figure 6-1 is organized in terms of the end product and that the lower level packages are items, whereas Figure 6-2 names

The convention of naming tasks with a verb and a noun is helpful.

work packages with the verb-noun format. We will leave it up to you and your organization to select the most helpful format.

Appropriate Level of Detail

Recall that the work packages are the unit of planning, execution, and control. Thus, you should size the work packages according to the desired level of control. There is no sense in having a small unit of work if the project cost accounting systems cannot provide data at that level of granularity or if the project manager will not pay attention to the detail.

On the other hand, there is an old saying, "The devil is in the details." To master the complexity of some projects, it is sometimes necessary to develop a WBS that might have hundreds and even thousands of work packages in it. At first, the quantity of work packages seems overwhelming, but team members soon recognize that this level of detail is important and even necessary.

Sometimes individuals and teams get into excessive detail in listing work packages, and the project plan begins to look more like a step-by-step procedure. They are well meaning in that they know that details often make the difference between success and failure, and they want to ensure that they (or someone else) do not forget anything. It is better to capture this detail in the WBS dictionary.

> **The size of the work packages depends on your control needs.**

A useful and widely recognized rule of thumb for sizing work packages is that each work package should contain approximately 40 to 80 hours of effort. The actual number and size of the work packages depend on a number of factors that stem from judgment and experience. The more work packages you have in your project, the smaller and cheaper each work package becomes. However, the more work packages you have, the more money and time are spent in arranging for these to be interfaced properly with each other and managed. As we discuss more fully in Chapter 17, small WBS tasks with short durations improve the precision of project status monitoring. Conversely, if you have only one work package, there is no interfacing cost, but the task itself is large and expensive. Another consideration in deciding how large a single WBS task should be is whether it will be the responsibility of a senior or junior person and their relevant experience. We summarize these considerations and other significant factors in Table 6-1.

VALIDATING THE WORK SCOPE

As we described in Chapter 5, project planning is a process of developing a "model" of the project. Planning is a dynamic activity. The composition and organization of the WBS is elaborated progressively as the project team strives to understand, describe, and estimate the work activities involved in the project. In practice, the team's development of the WBS is tightly coupled with its processes of estimating resources, cost, duration, and risk.

TABLE 6-1   Factors That Affect Task Size

Factor	When to Use Smaller Tasks	When to Use Larger Tasks
Management effort	You can afford to spend more time creating a WBS.	You want to spend less time creating a WBS.
Number of tasks	You want more detail.	You want less detail.
Spending authorization	You want to limit the amount of financial or resource commitment during a specific time interval.	Task managers have previously demonstrated prudent use of money and resources.
Task duration	You want to encourage faster completion of tasks.	You can wait longer to see tasks completed.
Monitoring accuracy	You desire or require greater accuracy.	You can tolerate less accuracy.
Company's prior experience with similar work	Your company has little or no previous experience.	Your company has expertise in specific work of the task(s).
Task manager's skill	You are using an inexperienced person.	You are using an experienced person.

If you can afford the time, it is desirable to have another person or group make a WBS for your project, independent of yours, decomposed to one or more granular levels of detail. This will highlight any discrepancies or oversights and may suggest a more effective way to organize the required work. You will have to repay the favor on later projects, but that should help your organization by reducing problems on projects. Some organizations require that two or more people independently prepare a WBS for a given customer-sponsored project before approval and baselining.

We suggest that you use the WBS and the organizational breakdown structure (OBS) to create a common graphic: the *responsibility-assignment matrix* (RAM). Figure 6-4 is an example of a RAM. You can see that each individual is assigned to one or more work packages. It is better to name individuals, but you also can list the organizational unit on the left side of the RAM.

Unclear roles and responsibilities are a frequent and significant complaint in projects. Here is a technique that is popular with some project managers. Notice in Figure 6-4 that there are letters inside the circles. RACI is an acronym for *r*esponsible, *a*ccountable, *c*onsulted, and *i*nformed. An individual who is responsible for a work package is the primary contributor to the work activity. When an individual is accountable for a work package, he or she typically is a functional manager who delegates the work to an employee who is the performer of the work activity. An individual who is consulted on a work package typically is a secondary contributor to the work activity. When an individual is informed on a work package, he or she is a person who "needs to know" about the work activity and frequently is a recipient of reports. (Figure 13-2 illustrates an alternative technique to clarify roles and responsibilities.)

Another method of validating the project work scope is to create a *cost breakdown structure* (CBS) that budgets the project cost to individual work packages.

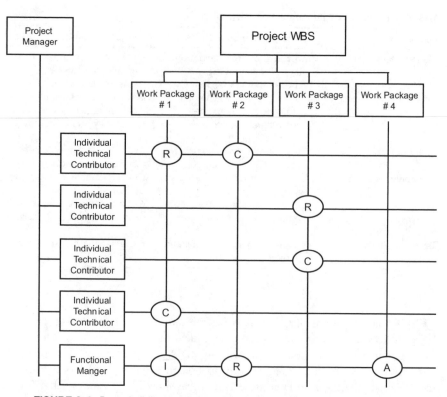

FIGURE 6-4. *Responsibility Assignment Matrix (RAM) showing RACI team roles.*

We will describe this concept more in Chapter 9. The project manager can validate both the WBS and the cost estimate through this approach and can evaluate performance risks in each work package at the given level of funding. This information helps the project manager to recognize the portions of the project that represent the most activity. Then, the project manager can make adjustments to balance effort, product performance, and project performance within the funding allocation.

The initial WBS is an input to schedule planning. The schedule planning may identify further items to add to the WBS. The WBS is *not* a schedule, and the project team constructs the WBS before it develops a project timeline. The WBS is a hierarchy of the work and does not show flow (or dependencies) between work packages.

As a final integration step, the project team updates the WBS to include these work packages so that everything on the WBS is finally tied to scheduled work packages and budgets, and vice versa. You should account for the phenomenon that some "larger" work packages will require further decomposition and should validate the cost and schedule against this, too. You are now at the point of establishing the project baseline, as described in Chapter 5. After baselining, the WBS does not change, except as permitted by the project's integrated change management process.

WORK SCOPE IS FUNDAMENTAL TO PROJECT INTEGRATION

This is a point worth emphasizing: A poor WBS undermines all project planning, execution, and control work. The WBS is an input to schedule development, risk management, cost estimating, and other project management concerns. The goal of project scope management, and the WBS, is to capture *all the work and only the work needed to produce the product of the project.*

The WBS is the central tool for project scope management.

As described in Chapter 1 and elsewhere, the project manager brings integration expertise. The benefits of the WBS as an integrative tool include

- *It defines work scope; thus, it defines the project.* The WBS defines all work and only the work that the project intends to do: Only the work captured in the WBS is in the scope of the project. You cannot estimate project duration or cost without some knowledge of the detail of what is in and out of the work scope. Organizations need the WBS as a basis of configuration management.

- *It is a basis for cost and scheduling.* You can't estimate something you can't define; thus, conventional project management thinking is that the work needs to be defined before any commitments to dates or budgets are accepted. The WBS provides a framework for decomposing the project into work packages so that the team thus can schedule and cost the packages.

The WBS helps to organize estimates and schedules.

- *It is a basis for executing work.* Work is managed at the work package level so that the project manager can verify the acceptability of work. The project manager controls the opening and closing of work packages. Consider the example shown in Figure 6-5, which clarifies the work for each task. Note that this form has a block where the task leader accepts the task. The simple form shown in Figure 6-5 is similar to what you would use if you subcontracted a task to another company. That is, it is a written agreement intended to get agreement on what is to be done, by when, and for how much (specified in labor hours and/or dollars). While it doesn't guarantee that the task will be completed as expected, it reduces the likelihood of misunderstanding between the requester and the doer. We will show you other roles for this form in Figures 17-3 and 20-1, demonstrating its substantial utility.

- *It is a basis for project cost accounting.* The project can capture and budget cost in each work package and the intermediate levels of detail. Each element

TASK AUTHORIZATION		PAGE OF
TITLE		
PROJECT NO.	**TASK NO.**	**DATE ISSUED**
STATEMENT OF WORK:		
APPLICABLE DOCUMENTS:		
SCHEDULE **START DATE:** **COMPLETION DATE:**		
COST:		
ORIGINATED BY: **DATE:**	**ACCEPTED BY:** **DATE:**	
APPROVED BY: **DATE:**	**APPROVED BY:** **DATE:**	
APPROVED BY: **DATE:**	**APPROVED BY:** **DATE:**	

FIGURE 6-5. Task authorization form.

of the WBS should be numbered (a standard convention is to use an outline such as 100.01.03, where the decimal points show the hierarchical relationships. Indeed, the numbering is one of the indicators of a good WBS because it allows tracing to cost, risk, and objectives. The customer in the case of some projects awarded by a contract, for instance, those to provide certain weapon systems to the Department of Defense, can dictate the WBS layout and numbering.

- *It is a basis for change control.* It is a fundamental preventative measure for scope creep, overengineering, and other common project ills. In general, it is best to structure the WBS for control reasons. Some people believe that the WBS should align exactly or nearly exactly with tangible, deliverable items, both software and hardware.
- *It is a basis for communicating responsibilities.* This provides the framework for subsequent delegation of effort. Many people use it as part of a role clarity

discussion (primary and shared responsibilities) for deliverables. These projects link team member goals with team goals and gain alignment of goals and incentives. The program manager generally does not get involved below the work package level unless a team member doesn't understand his or her job or is not motivated. The WBS is a useful tool to obtain cross-functional team involvement and commitment to the plan.

- *It is a basis for risk management.* Work packages provide inputs to risk identification as well as a way to assign responsibility to risk, and the WBS itself provides a framework for describing all project risk

HIGHLIGHTS

- A work breakdown structure (WBS) identifies all the work and only the work that the project will perform.
- Project scope management ensures that the project does all the work required, and only the work required, to complete the project successfully.
- The detail of the work package is determined by the control needs of the project
- A coworker's independently produced WBS for your project may identify omissions on your WBS. Use several iterations of level 1 of the WBS to ensure that you have a complete WBS.

7

Scheduling

This chapter and the next describe methods for developing and presenting project time schedules. In this chapter, we present three methods for displaying time schedules: bar charts, milestones, and network diagrams.

In Chapter 5, we discussed the general approach to integrated project planning, and we said that project planning is a team-based process of understanding the requirements and generating assumptions. In Chapter 6, we discussed the methods of identifying the work and organizing it, which is necessary prior to developing a schedule. Eventually, you will need to communicate your project model, and creating a presentation of the time schedule is one common element found in an integrated project plan.

OVERVIEW OF SCHEDULING FORMATS

Table 7-1 provides an overview of the common formats for displaying time schedules. Bar charts portray the time schedule of activities or tasks, and milestone charts portray the schedule of selected key events. Network diagrams portray activities, events, or both and explicitly depict their interdependency with predecessors and successors. Time-scaled tasks with explicit task interdependency linkage schedules (linked Gantt charts) portray both interdependency and the time sequence. The best choice for you may depend on what you are doing in a specific situation.

BAR CHARTS

Bar charts, often called *Gantt charts* after H. L. Gantt, an industrial engineer who popularized them during World War I, are a common format for displaying project schedules. Figure 7-1 is a bar chart. The project is divided into five activities with

TABLE 7-1   Comparison of Scheduling Approaches

Dependency Picture	Linear Time Scale	
	No	Yes
No	Lists of milestones or tasks	Bar (Gantt) charts or milestone charts

Yes	Network diagrams
	  Event-in-node (PERT)
	  Activity-in-node (PDM)
	  Activity-on-arrow (ADM)
	  Hybrids with anything in
	    node or on arrow
		Time-scaled tasks with explicit task
		  interdependency linkages (linked
		  Gantt)

a planned duration of 12 months. When the bar chart was constructed, five open bars were drawn to represent the planned time span for each activity. The figure also shows project status at the end of the sixth month. The shaded bars represent the forecasted span of the activities as of the end of the sixth month. Activity A was completed early. Activity B is forecast to be finished half a month late. Activity C is forecast to end approximately a month and a half early. The percentage of completion for each activity in the process is also illustrated. Activity A has been completed, B is 80 percent complete, and C is 30 percent complete.

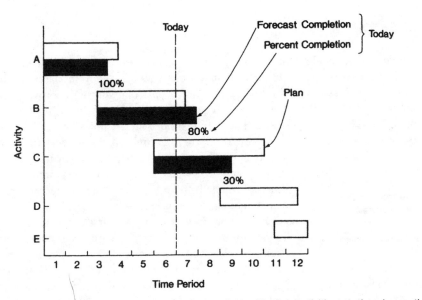

FIGURE 7-1. *Typical bar chart illustrating a project with five activities at the six-month review.*

Know the advantages and disadvantages of bar charts.

Bar charts are simple to construct and easy to understand and change. Because they show graphically which activities are ahead of or behind schedule, they are excellent communication tools for showing stakeholders baseline schedule data and actual schedule data. However, when compared with methods that force understanding of the predecessor-successor tasks, Gantt charts are weak planning tools.

Offsetting these favorable features are some weaknesses. One is that the notion of a percentage completion is difficult and is associated most commonly with the use of bar charts for measuring progress (which we discuss further in Part 4). Does the percentage completion refer to the performance dimension, the schedule dimension, or the cost dimension of the job? Unless an activity is linearly measurable, for instance, drilling a hundred holes in a steel plate, it is impossible to judge what percentage of it is complete. (Even in this simple case, the steel plate may have an internal defect, and the last drill hole might be through that defect, causing the plate to crack, at which point what was 99 percent complete now has to be done all over again.) Therefore, percentage completion becomes highly subjective or frequently is taken merely as the percentage of cost expended compared with total projected cost. In neither situation is percentage completion a useful number. Bar charts are much more useful as an indication of what has happened than as a planning tool to aid the project manager in making things happen properly in the future.

MILESTONES

A milestone schedule notes a few key events, called *milestones,* on a calendar bar chart. Milestones have been defined in various ways, but they probably are best defined as events clearly verifiable by other people or requiring approval before proceeding further. If milestones are so defined, projects normally will not have so many that the conclusion of each activity itself becomes a milestone.

A schedule that does not show task or activity interdependencies is useless by itself for planning.

The key to helpful use of milestones is selectivity. If you use only a few key events—perhaps one every three months or so—you will avoid turning milestones into pebbles (sometimes called *inchstones*) over which people are always stumbling. Some useful milestones might be, for instance, a major design review or a first article test.

When milestones have been defined, for instance, in the customer's request for a proposal (RFP) or in your proposal document, listing them often helps in preparing your project plan. Having such milestones with attendant schedule and budget measures adds extra emphasis to a few key points of a project. In common with bar charts, however, milestone schedules do not clarify activity or task interdependencies.

NETWORK DIAGRAMS

Network diagram is a generic term for what is also known as a *precedence diagram.* Over the years, these variants of network diagrams have been called the *precedence diagramming method* (PDM), the *arrow diagramming method* (ADM), the *program evaluation and review technique* (PERT), *bubble diagrams,* and many others. The following list contains some of these and their common abbreviations (that we will use):

PERT   Event-in-node (EIN)   Program evaluation and review technique
PDM   Activity-in-node (AIN)   Precedence diagramming method
ADM   Activity-on-arrow (AOA)   Arrow diagramming method
TBAOA   AOA with linear time scale   Time-based AOA

Network diagrams are recommended for planning the project schedule dimension. They identify the precedence conditions and the sequential constraints for each activity.

PERT and ADM emerged in different ways in the late 1950s. PERT is event-oriented (i.e., the event labels go in the nodes of the diagram) and typically has been used for aerospace and research and development (R&D) projects for which the time for each activity is uncertain. ADM is activity-oriented (i.e., the activity labels are placed on the arrows) and has been applied to the construction industry, in which there is typically a controllable time for each activity. Figure 7-2 portrays these two forms, as well as activity-in-node, most commonly called a *precedence diagram.* (As we said before, hybrids are common today primarily owing to computer-based project management software packages.) For instance, an activity-in-node may be zero time, namely, a milestone.

THE NETWORK LOGIC DIAGRAM

A *network logic diagram* is any of several displays that link project activities (or tasks) and events with one another to portray interdependencies. A single activity

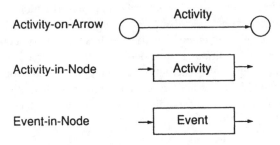

FIGURE 7-2. *Three types of network diagrams.*

or event may have interdependencies with predecessor, successor, and parallel activities or events. Figure 7-3 compares the three principal types: activity-in-node (AIN), activity-on-arrow (AOA, also called task-on-arrow), and event-in-node (EIN).

One of the few useful things a project manager can do when the project is in some difficulty is to change the allocation of resources dedicated to an activity. Consequently, if you are not already wedded to and successful with the EIN format, adopt one of the other two forms that explicitly display activities.

Linked Gantt charts depict both interdependency *and* the time schedule. People familiar with bar charts find this easier to understand and apply.

Conventions

Arrows designate activities or tasks.

Figure 7-4 shows how an amplifier building project would be illustrated in EIN and AOA diagrams. Figure 7-5 provides the symbolic conventions common to both. Activities conventionally are shown as arrows, with the start being the tail of the arrow and completion being the barb. Events are shown as circles (or squares, ovals, or any other convenient closed figure). The event number is placed inside the closed figure. Event numbers occasionally are required in computer programs, which sometimes are used to facilitate network information manipulation. In such cases, a higher number activity always follows a lower number activity. In some computer programs, activities are not labeled by their name (i.e., "build amplifier" is not called "build amplifier") but rather by the start and end numbers (i.e., 5-10 is build the amplifier if 5 is the number of the start node and 10 is the number of the finish node). It is also conventional to place early and late date numbers and other information such as slack (or float) time within the nodes. Whenever node numbers are used, there should be a legend explaining which number is which. Using these conventions, a network diagram consists of a series of nodes and arrows connected to show the order of activities.

A dummy activity is a precedence condition.

The upper drawing of Figure 7-6 depicts an AOA schedule plan in which activity R must be complete before activity S can commence and in which activity T must be complete before activity U can commence. The middle drawing shows a schedule plan in which both activities R and T must be complete before either S or U can commence. The bottom drawing introduces the concept of a *dummy activity,* which is an activity requiring no work, that is, a precedence condition. It thus depicts a plan in which both activities R and T must be complete before activity S can commerce and in which activity T cells must be complete before activity U can commence. Activity U does not depend on activity R, because the dummy arrow points in the other direction.

Network Terms

Figure 7-7 illustrates three terms in network diagram usage. A *burst node* (*node 2*) is a node or an event at which two or more activities can be initiated after com-

ON ARROW

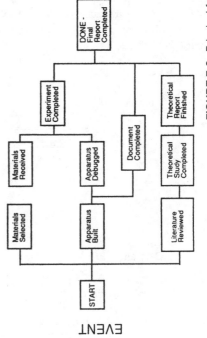

IN NODE

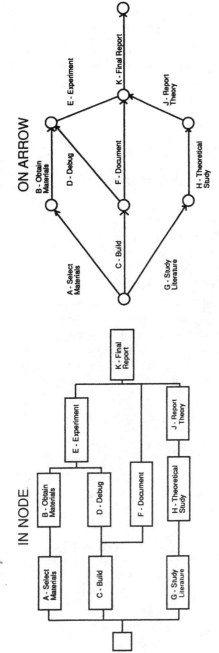

FIGURE 7-3. Principal forms of network diagrams.

EIN (Events Labeled)

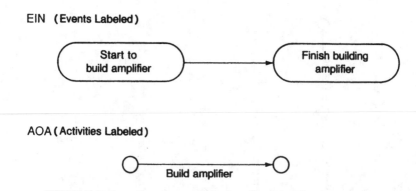

AOA (Activities Labeled)

FIGURE 7-4. *Examples of event-in-node and activity-on-arrow diagrams.*

pletion of a preceding activity. A *merge node* or *event* (node 5) is one in which two or more activities must be completed prior to initiation of the subsequent activity.

A *dummy activity* represents a dependency between two activities for which no work is specifically required. Dummies are also used to deal with an ambiguity that arises in some computer-based network diagrams, also illustrated in Figure 7-7. As mentioned, in computer-based network diagram programs, the activity label sometimes is not the activity name but rather the number of the two nodes preceding and following it. Thus, is activity 6–7 task U or task H in the lower left of Figure 7-7? Using a dummy task, one can make activity 6–7 task G and activity

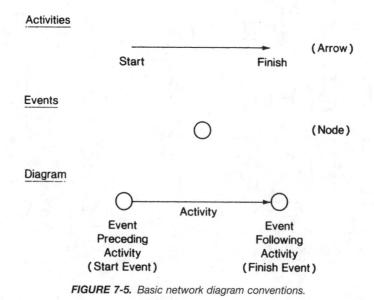

FIGURE 7-5. *Basic network diagram conventions.*

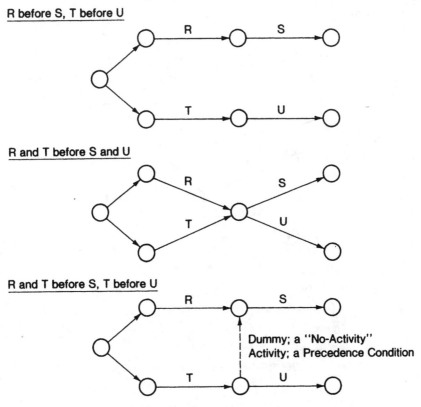

R before S, T before U

R and T before S and U

R and T before S, T before U

Dummy; a "No-Activity"
Activity; a Precedence Condition

FIGURE 7-6. Precedence requirements.

6–8 task H, as shown in the lower right of Figure 7-7. There is thus a dummy activity, 7–8, also required to remove the previous ambiguity.

WHY USE A NETWORK DIAGRAM?

An Illustrative Situation

Consider the project situation illustrated by the bar chart in Figure 7-8. You are the project manager for this project with eight activities or tasks. At the end of four months, you are conducting a project review (denoted by the triangle). Task managers provide status reports (denoted by the shading) showing that tasks B and C are two months late, A and D are one month late, and E is on schedule. The impact of these delays on the entire project's completion is not clear. (For simplicity in this example, assume that task status is precisely measurable—for instance, by counting the number of holes drilled or drawings completed.) Your chief concern now is whether the entire project is late.

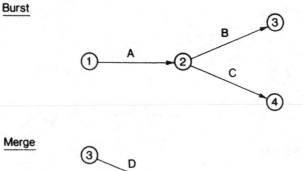

Burst

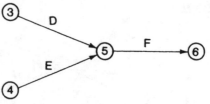

Merge

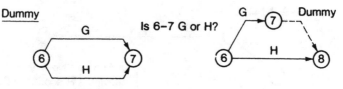

Dummy

Is 6–7 G or H?

FIGURE 7-7. *Common network terms.*

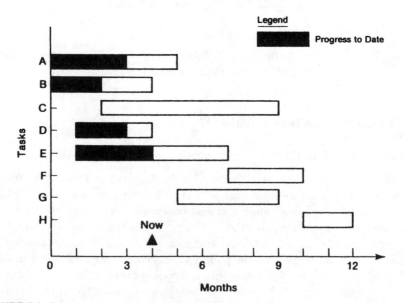

FIGURE 7-8. *Bar chart showing tasks A, B, C, and D to be late at four-month project review.*

First, you might break tasks into their subtasks or subactivities, as in Figure 7-9. This provides additional information but still does not tell us whether the project will be late.

An Event-in-Node Network Diagram   At this point, a network diagram for the project can be examined (Figure 7-10). This contains more information than the bar charts because it shows the interrelationships (precedence) between different tasks. We have labeled each event in its node with the completion of the designated subtask. Thus, at the top of the diagram, you can see that activity D must be completed before activity G.

Networks have more information than bar charts.

Figure 7-10 also shows the problem with an EIN network. The activities per se are not illustrated; that is, there is no arrow uniquely associated with activities G, H, or C_3. This will always be the case where two or more arrows come to a single node, that is, at all merge nodes. This is not a problem for a skilled practitioner, but it does seem to present an unnecessary conceptual difficulty. When the activities are not shown explicitly on a diagram, it can be more difficult for the project manager and others to visualize them and their relationship.

A project manager's ability to influence the course of his or her project depends on his or her ability to influence the work on a given task or activity. One of the

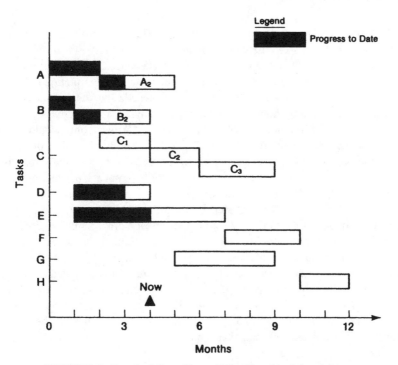

FIGURE 7-9. Bar chart (from Figure 7-8) with subtask breakdown.

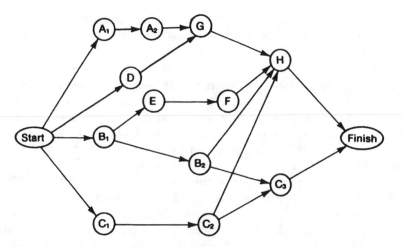

FIGURE 7-10. *Event-in-node network diagram for the project in Figure 7-9. Each node designates completion of the indicated task or subtask.*

few things a project manager can do is change the allocation of resources devoted to a particular activity. Thus, the lack of each activity's explicit visibility in an EIN diagram may be troubling.

An Activity-On-Arrow Network Diagram   This EIN problem can be eliminated if we go to an AOA network (Figure 7-11), which shows all activities by labels on the arrows. It clearly indicates the precedences. The requirement of a dummy activity, a "no activity" activity, is to indicate that completion of activities B_2 and C_2 (as well as activities F and G) must precede the start of activity H. In this AOA representation, though, the merge nodes are not single-activity completion events. For instance, the node to which the D and A_2 arrows come would not have to be designated the completion of both D and A_2. Frankly, this AOA duality does not trouble one of us (MDR) as much as the EIN's lack of activity emphasis and clarity. In fact, the duality actually may be helpful by emphasizing that both D and A_2 must be completed before G can commerce.

Figure 7-12 illustrates the next step in using the AOA diagram: redrawing it to a time scale in which the horizontal projection of each arrow is proportional to the amount of time required for its activity. This is essentially a linked Gantt chart diagram, in which the horizontal distance spanned by an arrow is the time schedule, and the predecessors and successors for each task are clearly revealed by junctions. Doing this reveals that one path (B_1, E, F, H) is longer than any other. This is called the *critical path*. It also may be identified as the path that contains no *slack time* (the amount of time available on a path that is the difference between that required on the critical path and that required on the particular activity path with slack time).

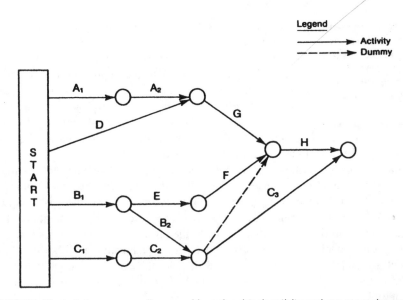

FIGURE 7-11. *Activity-on-arrow diagram with each subtask activity and one precedence condition (or dummy).*

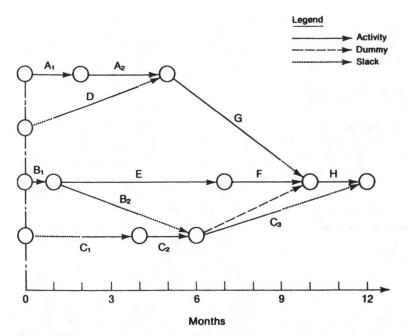

Months

FIGURE 7-12. *Time-based linked Gantt chart drawn on assumption of the task schedules shown in Figure 7-9. (Note that start node representation is an alternative to that used in Figure 7-9.)*

Figure 7-12 is drawn with each activity shown, starting at the time it was scheduled in Figure 7-9. Noting the task progress status and current date, Figure 7-13 begins to provide direct information as to the implications of the delay on activities A, B, C, and D. Figure 7-14 is a redrawn version of Figure 7-13 in which the project manager has taken advantage of the slack time. That is, all late (delayed) activities are drawn to show the work remaining to be done, and subsequent tasks thus are rescheduled in several cases.

Thus, although activities A, B, C, and D are in fact later than planned, the project has not yet suffered any irretrievable schedule slippage. But the project now has two critical paths, whereas it previously had only one. That is, there is no longer any slack in the upper branch (A_2 and G). Because there is still one month of slack on the lower path (task C), perhaps some of the resources allocated to it (or to task B_2) might be redeployed to one of the other critical-path activities. It is vastly more difficult to complete a project on time with more than one critical path, and it is unlikely that this project will be completed on schedule, although it is not yet irretrievably lost.

Another Example

Figures 7-15 and 7-16 are an AOA diagram and two versions of bar charts for a house-painting project. The network diagram clearly contains far more information than either of the bar charts—for instance, the dependency of D_1 and D_2 on A_3.

COMPUTER SOFTWARE

There is no assurance that a project management computer software package will produce exactly the format you desire. Many produce box displays with both activity and events in the boxes (or nodes), and this may be called *PERT.* Figure 7-17 is an example, which shows precedence but is not linear in time. Figure 7-18 is the linked Gantt chart, and Figure 7-19 is the simple Gantt chart. All three of these charts are derived from the same input data for task names, durations, and predecessor tasks.

Another useful feature of project management software is the ability to link tasks or activities in several different ways, as illustrated in Figure 7-20. Each of these time linkages may be zero, plus, or minus time in the most flexible software. We commonly use finish to start with zero time separation, as illustrated in Figure 7-9. However, for a home-redecorating project, if task one is to shampoo the carpet and task two is to move in the furniture, you might wish to have plus two days for the linkage to allow the carpet to dry. No work is required during those two days, but it does affect the overall schedule. Similarly, in the start-to-start situation,

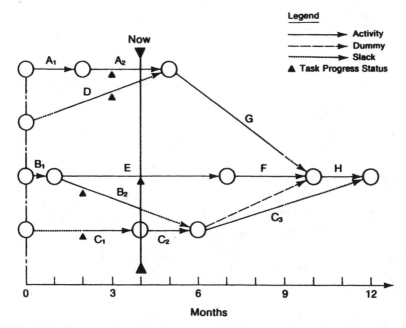

FIGURE 7-13. *Figure 7-12 with task progress status from Figure 7-9 denoted by solid triangles.*

task three might be to perform the test, and task four might be to prepare the test report. Task four can start a few days after task three has started and need not wait for the completion of task three for its initiation.

HELPFUL HINTS

One frequently asked question is, "How do I start a network diagram?" One answer is, "With lots of scrap paper." But the best way to start is with the WBS. From the WBS, you can start the network diagram from either the beginning or the end of the project. There are frequently somewhat obvious large subnets you can quickly put down on a piece

Include every element in the WBS in the network diagram.

of scrap paper. As a general rule, it is probably best to start from each end with scrap paper and sort out the connectedness in the diagram, where there are activities in progress simultaneously. Another "paper and pencil" technique is to write the name of each task on a sticky note, perhaps using different colors for different groups' responsibilities. These notes then can be arranged on a large roll of paper in time sequence and these results transferred either to a master schedule or into computer software. As we said in Chapter 6, you may now find some work tasks that should be inserted into the WBS. If so, revise the WBS accordingly.

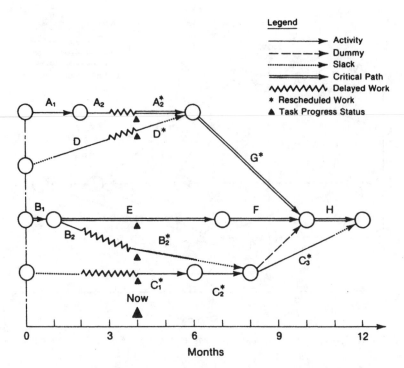

Legend

→ Activity
⇢ Dummy
⋯⋯▶ Slack
⇒ Critical Path
∿∿∿ Delayed Work
∗ Rescheduled Work
▲ Task Progress Status

FIGURE 7-14. *Time-based linked Gantt chart redrawn to show delayed work and rescheduled activities, permitting the project to be completed in accordance with the original 12-month schedule.*

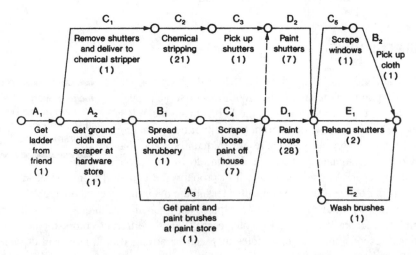

FIGURE 7-15. *Activity-on-arrow networks for house-painting project, with activity duration in days.*

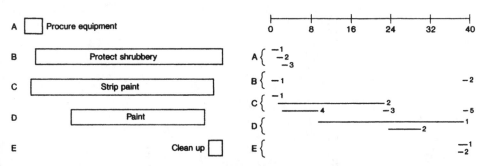

FIGURE 7-16. Bar charts for house-painting project, again revealing lack of task interdependency information.

You then can transfer the entire diagram to a clean piece of paper. It is probably helpful to do this with a time base and with the presumption that each activity starts at the earliest possible time. For this initial version, we recommend that you get the people who will be responsible for each activity to estimate how long it will take to carry it out on a normal work basis. When you put these time estimates onto the network diagram, it may become apparent that the entire project will take too long. At this point, you can identify particular activities that may be candidates for time compression, that is, tasks that you believe can be done faster.

As shown in Figure 7-21, network diagrams may require crossover of lines. This is to be expected. Although some diagram rearrangement may get rid of crossovers, it also may distort the logical relationship of groups of activities, for instance, all those being carried out by one department being within a general band of the diagram. If activities A, B, C, D, E, F, and G in Figure 7-21 are performed by one section, the upper diagram, which has two crossover intersections, would be preferable to the lower one. Thus, there are cases in which increased use of crossovers will be clearer. Crossovers in some computer project management software can be especially confusing, and there may not be much you can do about it.

The project manager should construct a network diagram of perhaps three dozen activities or up to five dozen, if required. Such a diagram normally can be drawn in less than two hours and will fit onto a standard 17- by 22-inch sheet of paper. If some activity in this network is very large, its activity manager can make a network diagram for it. In this way, with a few hand-drawn networks of a few dozen activities each, large projects can be handled without the use of a computer-based network system. Given continuing improvements to and the flexibility of computer project management software, today it may be easier to use software than manual techniques. However, in common with most computer software, project management software improvements and upgrades often are accompanied by a bewildering increase in program features and complexity that may be unnecessary for your needs.

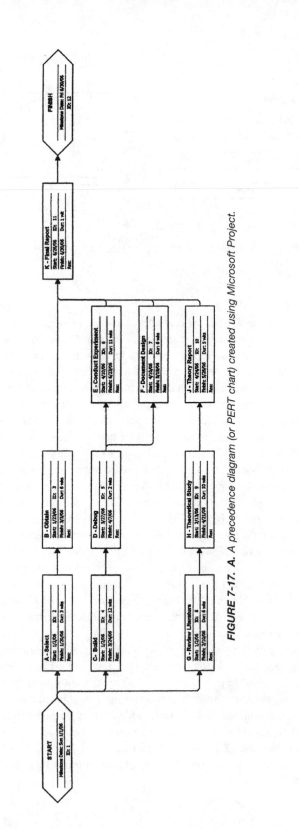

FIGURE 7-17. A. A precedence diagram (or PERT chart) created using Microsoft Project.

116

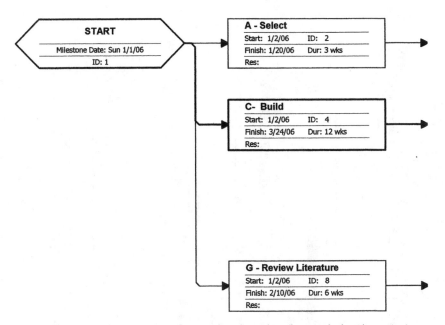

FIGURE 7-17. B. *Initial nodes for part A, enlarged to show typical node content.*

Some project (and other) managers resist the use of network diagrams because they consider them complex. This is a serious mistake. It is not the network diagram that is complex; it is the project itself that is complex. In fact, if you can't draw a network diagram for your project, that should be a clear danger signal that you do not understand your project. An advantage of using network logic diagrams is that you are forced to confront the logic of your entire project. There is an obvious problem if you cannot load data to get a complete solution, assuming that you or a member of your project team is able to use a computer.

If you can't make a network diagram, you can't run the project.

As we wrote in Chapter 5, it is generally better to build a paper-and-pencil model before going to the computer. We are not opposed to the use of a computer-based planning network program to assist with the mechanics of network usage. In fact, computer programs have great value in determining resource requirements quickly. However, a computer-based network diagram program is not required to manage many projects, and the lack of such computer assistance is no excuse for not using a linked Gantt chart or precedence diagram.

Always use a network diagram to plan the schedule dimension, even if you do not show it to your management or the customer.

Very large projects, however, normally will require a computer-based system, and many are available. However, you usually lose the "hands on" feel you can get by drawing your own network diagram. In general, avoid the computer until you have outlined the project on paper or a planning roll.

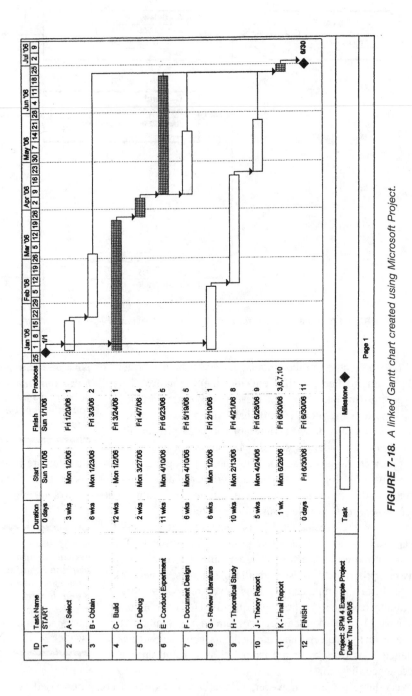

ID	Task Name	Duration	Start	Finish	Predeces
1	START	0 days	Sun 1/1/06	Sun 1/1/06	
2	A - Select	3 wks	Mon 1/2/06	Fri 1/20/06	1
3	B - Obtain	6 wks	Mon 1/23/06	Fri 3/3/06	2
4	C - Build	12 wks	Mon 1/2/06	Fri 3/24/06	1
5	D - Debug	2 wks	Mon 3/27/06	Fri 4/7/06	4
6	E - Conduct Experiment	11 wks	Mon 4/10/06	Fri 6/23/06	5
7	F - Document Design	6 wks	Mon 4/10/06	Fri 5/19/06	5
8	G - Review Literature	6 wks	Mon 1/2/06	Fri 2/10/06	1
9	H - Theoretical Study	10 wks	Mon 2/13/06	Fri 4/21/06	8
10	J - Theory Report	5 wks	Mon 4/24/06	Fri 5/26/06	9
11	K - Final Report	1 wk	Mon 6/26/06	Fri 6/30/06	3,6,7,10
12	FINISH	0 days	Fri 6/30/06	Fri 6/30/06	11

Project: SPM 4 Example Project
Date: Thu 10/6/05

Task ▭   Milestone ◆

Page 1

FIGURE 7-18. A linked Gantt chart created using Microsoft Project.

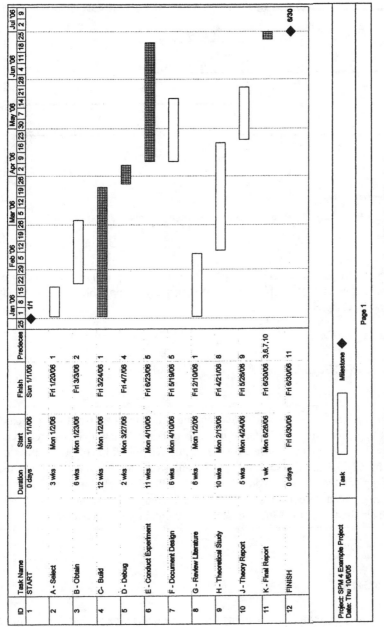

ID	Task Name	Duration	Start	Finish	Predeces
1	START	0 days	Sun 1/1/06	Sun 1/1/06	
2	A - Select	3 wks	Mon 1/2/06	Fri 1/20/06	1
3	B - Obtain	6 wks	Mon 1/23/06	Fri 3/3/06	2
4	C- Build	12 wks	Mon 1/2/06	Fri 3/24/06	1
5	D - Debug	2 wks	Mon 3/27/06	Fri 4/7/06	4
6	E - Conduct Experiment	11 wks	Mon 4/10/06	Fri 6/23/06	5
7	F - Document Design	6 wks	Mon 4/10/06	Fri 5/19/06	5
8	G - Review Literature	6 wks	Mon 1/2/06	Fri 2/10/06	1
9	H - Theoretical Study	10 wks	Mon 2/13/06	Fri 4/21/06	8
10	J - Theory Report	5 wks	Mon 4/24/06	Fri 5/26/06	9
11	K - Final Report	1 wk	Mon 6/26/06	Fri 6/30/06	3,6,7,10
12	FINISH	0 days	Fri 6/30/06	Fri 6/30/06	11

Project: SPM 4 Example Project
Date: Thu 10/6/05

Task [ ]   Milestone ◆

Page 1

FIGURE 7-19. A bar chart layout created using Microsoft Project.

119

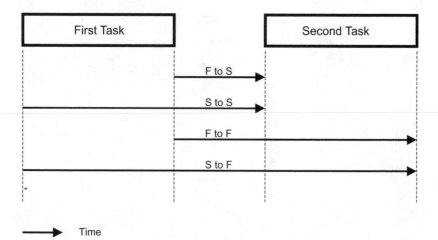

FIGURE 7-20. *Possible linkages between task start (S) and finish (F).*

Once you have the relationships between activities identified, you can estimate the time of each activity as accurately as possible using techniques described in Chapter 8.

TYPICAL PROBLEMS

In many ways, the worst schedule planning problem is to avoid the indicated scheduling problems. For instance, the schedule may show that required materials will not arrive at the needed time. Some people will avoid or dismiss the issue by saying that the schedule can be adjusted later. Maybe it can, but that is hoping for luck to save your project schedule. The solution is to admit the problem exists and revise the schedule to overcome it—now, not when there is no longer time to correct the problem and maintain your schedule.

HIGHLIGHTS

- Network diagrams show activity interdependencies, and common forms are PDM and ADM.
- A network logic diagram is essential for project planning because it shows the relationships between all the work packages or activities in the WBS.

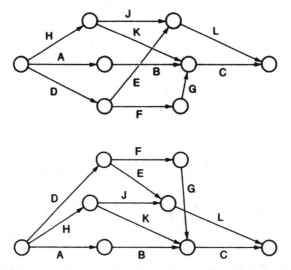

FIGURE 7-21. *Activity-on-arrow network diagram with crossovers.*

- It is better to think of Gantt (bar) charts and milestone charts as communication tools intended for audiences outside the project. If you develop a good network logic diagram, it is relatively easy to develop the Gantt chart or milestone chart.
- Consider the advantages of a time-scaled task with explicit task interdependency linkages (linked Gantt) format.

8

Time Estimating and Compressing the Schedule

This chapter continues the previous chapter's discussion of scheduling techniques, focusing on time estimating and speeding up project completion.

TYPES OF TIME ESTIMATES

Time estimates inherently have some error and thus are inaccurate to some extent. Your goal is to understand the degree of error.

Labor hours plus nonlabor costs and task duration all must be estimated.

A schedule for any project requires an estimate of the duration for each activity or task. Because, by definition, the project is unique, there is an inaccuracy in such time estimates. (The only way to guarantee meeting a time estimate is to make it infinitely long, in which case the project won't be authorized.)

After developing the work breakdown structure (WBS) and milestones that are necessary to generate a project plan, you need to assess their accuracy. How do you judge if the component estimates are ambitious, reasonable, or overly conservative? Although some projects can be estimated based on similarity to past efforts, many are radically different or require development of new technology or processing methods and thus are hard to estimate accurately. The best you can do is strive to be reasonably accurate in your time estimates. If you estimate a modestly large number of tasks this way, there will be compensating over- and underestimates of time (and cost). The total project estimate will have a smaller percentage error if your team makes estimates for a larger number of tasks, assuming random small over- and underestimate errors for individual tasks. If some tasks on your project are identical or very similar to tasks performed previously, it is easier to estimate these.

(As we said in Chapter 6, one of the goals of breaking the project into small work packages is to obtain understandable tasks, which also implies that they can be estimated.) There are two time-estimating techniques you may wish to use: PERT and pragmatic. Keep in mind, however, that there are two issues in time estimating. The first is to establish the number of labor hours required for a given task (which may depend on the specific skill levels of the personnel). The second is to determine the time duration for this activity. You must know the former for cost planning and managing the project, and the latter will determine the overall project schedule. If your schedule depends on the availability of a needed resource that may not be available when required, you should document that as an estimating assumption and include it in your issues management and risk approach.

Don't Pad Task Durations and Cost

Padding is a form of risk acceptance that involves expanding a task's duration or level of effort to protect for unknowns. Some people call it "adding fat." Unfortunately, padding is a common practice, but increasingly organizations recognize and discourage its use. Here are some of the reasons:

- It is hard to "back out of" the estimate to understand the amount of extra resources inserted. Therefore, the project manager has a distorted baseline to compare actual performance against. Conceptually, the practice of padding confuses two distinct project management activities: estimating and budgeting.
- Work typically expands to fill the time allowed; thus, work is more likely to come in over the estimate than under the estimate. Therefore, control of the project shifts away from the project manager to individual performers.
- If one person pads more than another, it could lead to a misidentified critical path.
- People spend increasing energy "gaming" their estimates to a number rather than probing for validity of assumptions. Eventually, no one trusts any estimate, and the culture becomes more cynical.
- If you submit knowingly erroneous (false) estimates on certain kinds of government contracts, you expose your company and yourself to potential felony charges. Make sure that you understand your company's stated policies on estimating and risk reserves before you submit an estimate to the government.

PERT Estimating

The program evaluation review technique (PERT) originated in projects characterized by uncertain times for activities. This problem was dealt with by requiring three time estimates for each activity:

1. The most likely (or probable) activity time (T_m), considering resource commitments or prospects that allow for schedule dependencies and probable conflicts or interruptions.
2. The optimistic activity time (T_o), namely, the shortest time that might be achieved under best-case conditions for the planning assumptions in the most likely case.
3. The pessimistic activity time (T_p), namely, the longest time that might be achieved under worst-case conditions for the planning assumptions in the most likely case.

As Figure 8-1 shows, this permits calculation of the expected time for the activity (T_e). PERT estimating is a reasonable way to estimate. You also can calculate the uncertainty of that time, which is called the *standard deviation* (σ). The calculation is illustrated in Figure 8-2.

PERT time estimating is useful when the time schedule is critical.

Figure 8-3 shows how to figure the expected time for a path and standard deviation of the path's expected time. The significance of the calculated standard deviation is the same as with the normal (Gaussian) probability distribution: Two-thirds of the time, the work will be completed within plus or minus one standard deviation of the expected time; 95 percent of the time, it will be completed within two standard deviations; and 99 percent of the time, it will be completed within three standard deviations. This kind of calculation can be important and helpful if there will be a cost penalty for lateness because you can estimate the likelihood of being late.

Figure 8-4 illustrates the three time estimates for a particular path containing three activities. Completion of the calculations for this case would show that the expected time is 29 days, and the standard deviation is 6 days. Therefore, the project's completion would be between the twenty-third and thirty-fifth day two-thirds of the time.

Many people have found it a helpful practice to ask others for three point estimates (optimistic, pessimistic, and most likely) and then follow up with

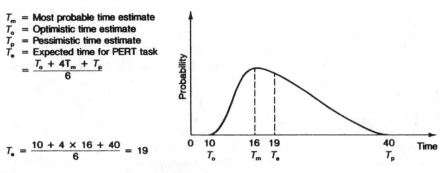

T_m = Most probable time estimate
T_o = Optimistic time estimate
T_p = Pessimistic time estimate
T_e = Expected time for PERT task
$$= \frac{T_o + 4T_m + T_p}{6}$$

$$T_e = \frac{10 + 4 \times 16 + 40}{6} = 19$$

FIGURE 8-1. PERT time estimating.

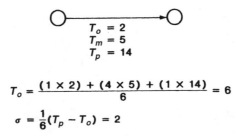

$$T_o = 2$$
$$T_m = 5$$
$$T_p = 14$$

$$T_o = \frac{(1 \times 2) + (4 \times 5) + (1 \times 14)}{6} = 6$$

$$\sigma = \frac{1}{6}(T_p - T_o) = 2$$

FIGURE 8-2. *PERT time uncertainty (σ-standard deviation) for a single event.*

a request for more information that is in this format: "Help me understand why this is an optimistic point estimate. What are your underlying assumptions?" This style of inquiry makes it clear to the other person that you are asking for help, not interrogating them. People who have used this approach have found that they obtain better information and are able to maintain a productive work relationship with the other person or group.

It requires some effort to make three time estimates and calculate the expected time and standard deviations. Many of the project management software packages can perform these PERT time estimate calculations.

Time Estimating

We recommend a collective approach for time estimating. The task leader, project manager, and one to three others should discuss the task and arrive at a judgment as to what the schedule should be. The task leader is there because he or she may have some first-hand knowledge of the work and will have to commit to the estimates. The project manager is there to provide balance with other project time estimates. The others are there to bring expertise and experience to bear.

As a practical matter, the project manager, task leader, and one to three others cannot hold discussions on every task on a large project because there simply is not enough time. In such a large project, this is the goal at which to aim, and a compromise is for the project manager to have several deputies to represent him or her in task estimating meetings.

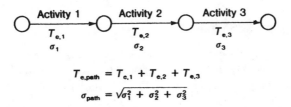

$$T_{e,path} = T_{e,1} + T_{e,2} + T_{e,3}$$

$$\sigma_{path} = \sqrt{\sigma_1^2 + \sigma_2^2 + \sigma_3^2}$$

FIGURE 8-3. *PERT expected time and uncertainty for a path.*

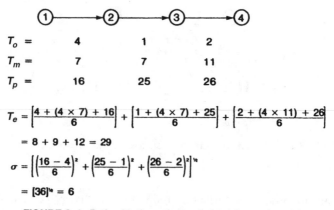

FIGURE 8-4. Path with three time estimates for each activity.

Base the time estimate for a task on who will do the work and how they will do it.

The goal is to establish and validate assumptions and then arrive at a consensus for the task's duration and resource load. If the group assumes that the task performer is a junior person, then it probably will take longer to complete the task compared with a senior person (who may be one of the consulted experts). Sometimes the reverse is true—a junior mechanical engineer may be able to complete a design very quickly using computer-aided design tools that older engineers have never learned to use.

If a task, as distinct from the project of which it is a part, is identical (or very similar) to a previously completed task, then the experience on the prior task is a good estimating guide. However, be certain that there are no meaningful differences that invalidate the relevancy of previous experience.

A logical sequence for estimating a new task is (1) to determine how many days the previous, similar task required and how many personnel worked on it by consulting existing company project records, (2) to decide how much more complex the present project is to arrive at a time duration and personnel multiplier, and (3) to determine the cost of the new task by multiplying the person-days by the appropriate labor rates. This assumes that such records exist and underscores the importance of retaining project histories. If no such records are retained, then memory is all that can be used, and memory may differ from one person to another.

Compressing the Schedule

As we pointed out in Chapter 3, project managers feel extraordinary pressure to complete projects quickly, and the entire project team needs to understand the issues involved in balancing competing demands. The initial critical-path duration will tell you whether or not you can meet the date targets.

The sensible approach is to replan tasks so that there is a credible reason to believe that they will be completed in times shorter than the first estimate. In general, start with tasks that are early in the project, and replan enough of these to achieve a satisfactory completion date. (Keep later project tasks at their first time estimate in case you subsequently need to recover from unplanned problems.) In doing this, try to compress tasks that have low compression costs and risks, and be careful not to create another critical path. If this will not save enough time, you

When a shorter schedule is required, select from the many proven project management techniques for schedule compression.

will have to replan the project entirely, perhaps running some tasks in parallel. Figure 8-5 shows that other options to shorten the schedule, if that is required, include negotiating for a simpler specification, using more effective resources, and devising a more efficient project plan.

In integrated project management, we work to maintain the logic and integrity of the project model. The operative rule is this: Schedule logically, and improve systematically. If you follow this rule, then there are several well known project management tools that can assist you in accelerating the project:

- *Assumptions analysis.* Often, initial assumptions are more conservative. As the project team better understands its project model, it can find satisfactory change assumptions that each individual can "buy into." This is often done as part of the project risk management practice.

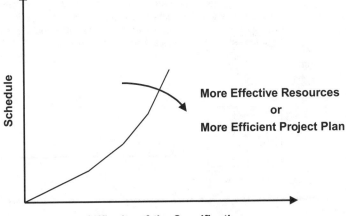

Difficulty of the Specification

FIGURE 8-5. More effective resources or a more efficient project plan can shorten the completing schedule for a given specification. See also Figure 2-2.

- *Fast tracking.* Fast tracking is the practice of overlapping phases and tasks, allowing work that initially was assumed as sequential to be performed concurrently with other activities. There may be increased risk in such a schedule, but that may be the lesser evil.
- *Crashing.* As described in Chapter 3, crashing is applying more resources to a project in order to speed it up
- *Simplifying.* Often, you can negotiate with the customer to remove requirements or to defer them

EARLIEST AND LATEST START AND FINISH TIMES

To introduce this important benefit of network diagrams, consider Figure 8-6. Although the activity durations are shown on the arrows, this is an EIN version, with time emphasis on the nodes. The name of the technique for calculating the early and late starts and finishes is the *forward-pass, backward-pass method.* Let's assume that the project starts at time equal to zero. The earliest time (T_E) you can emerge from the start node is zero. On the critical path, adding the activity time (in this case, 10), the earliest time you can get to the finish node is 10. Latest and earliest are always the same on nodes for the critical path; hence, $T_L = 10$ at the finish node and $T_L = 0$ at the start node. Off the critical path, the earliest you can reach the upper node is the earliest you can leave the start node plus the activity time on that path (in this case, 2).

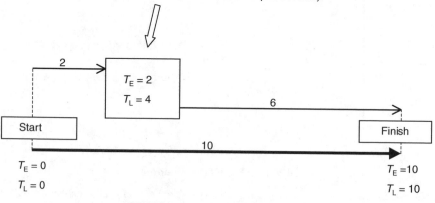

$T_E = T_L$ on the critical path. On other paths, $T_E = T_E$ of the prior node plus the activity time (or highest if two or more prior activities) and $T_L = T_L$ of the ending node minus the activity time (or lowest if two or more subsequent events).

FIGURE 8-6. Earliest and latest starts.

Latest times are calculated by working backward. Thus, T_L at the finish node (10) less the activity time (6) determines that the latest time you can leave the upper node (without delaying completion) is $T_L = 4$. The difference between $T_E = 2$ and $T_L = 4$ at this upper node (2) is the slack, or float, on the upper path.

The difference between earliest and latest times at a node indicates the amount of slack.

Now consider an AOA network diagram (Figure 8-7). The entire project is always assumed to start at time zero, Thus, the start of each activity that emerges from the start node has zero as its earliest start time (E_s). The earliest finish time (E_F) for each of these initial activities on the critical path is the duration of the activity itself (Figure 8-8) plus the earliest finish of the predecessor critical-path activity. For the entire network, earliest start is the duration of the activity itself (Figure 8-9A). Earliest start and finish times are calculated by proceeding from the start node to the finish node. In Figure 8-9A, activity duration is shown by the number above the middle of the activity arrows. The earliest finish of an activity is equal to the activity duration plus the earliest start. At the merge node, the earliest start of the following activity is the higher of the earliest finishes for the preceding activities. On the critical path, the earliest finish at the finish node is both the minimum project duration and the latest finish for that activity.

Figure 8-9B shows how to calculate latest finish (L_F) and latest start (L_S) times for each activity. Calculation commences at the finish node and proceeds backward to the start node. In Figure 8-9B, on the critical path, the latest and earliest times are equal. The latest start of an activity equals the latest finish of that activity minus

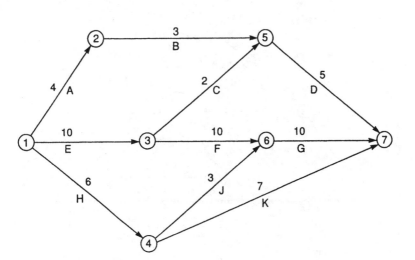

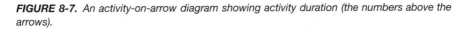

FIGURE 8-7. *An activity-on-arrow diagram showing activity duration (the numbers above the arrows).*

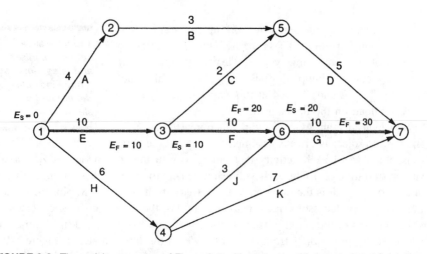

FIGURE 8-8. *The activity-on-arrow of Figure 8-7 with earliest and latest start and finish times on the critical path.*

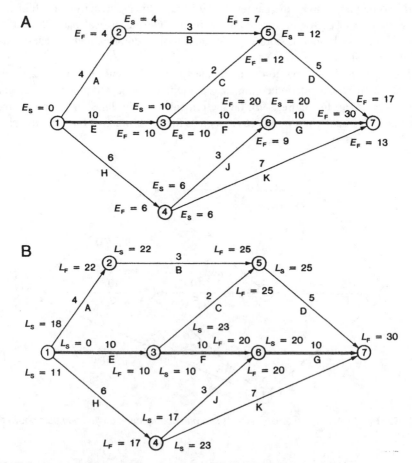

FIGURE 8-9. *(A) Earliest start (E_S) and earliest finish (E_F) calculations. (B) Latest start (L_S) and latest finish (L_F) calculations.*

TABLE 8-1   Typical Data Provided in a Computer Printout for Computer-Based Network Reporting (Use with Figure 8-9)

Event				Start		Finish		
Start	Finish	Description	Duration	E	L	E	L	Slack
1	2	Activity A	4	0	18	4	22	18
1	3	Activity E	10	0	0	10	10	0
1	4	Activity H	6	0	11	6	17	11
2	5	Activity B	3	4	22	7	25	18
3	5	Activity C	2	10	23	12	25	13
3	6	Activity F	10	10	10	20	20	0
4	6	Activity J	3	6	17	9	20	11
4	7	Activity K	7	6	23	13	30	17
5	7	Activity D	5	12	25	17	30	13
6	7	Activity G	10	20	20	30	30	0

the activity duration. At the burst nodes, the latest finish of the preceding activity is the lower of the latest starts for the following activities.

Table 8-1 shows the kinds of data provided by a typical computer printout for the project illustrated in Figure 8-9. Although less graphic, these data reveal the same information. Normally, all the earliest and latest information would be on one diagram. This is illustrated in Figure 8-10. Note that use of vertical dashed lines, without any dependency arrow indication, permits a node to be drawn in more than one location. This permits spatial separation of activities, thus providing additional open space on the network diagram.

TYPICAL PROBLEMS

There is an interaction of schedule with resources applied. A good designer is usually fast and accurate, and the later fabrication activities thus may be fairly quick. Conversely, a poor or junior draftsperson may be slower and less accurate, and subsequent activities may take longer. The plan reflects what you intend to do but is not necessarily what you will do. Another aspect of this resource interaction is that having two people on one task is not necessarily as productive as having one person taking twice as long because the two people must spend time communicating with each other.

Sometimes the project manager or higher management does not like the overall estimated length of the project and wants it reduced. Such a reduction is a problem if time is merely cut out of a task without changing the task's work plan to reflect how this reduction actually can be accomplished.

It is difficult to obtain accurate time estimates for things not done before. You also must note overly ambitious or risky approaches so that backups can be identified and planned in advance. As suggested earlier, getting a few people together, including especially those who will be responsible for the activity, and pooling judgments is the best solution to this problem.

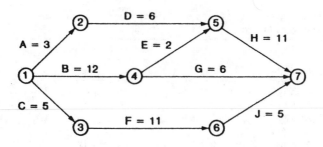

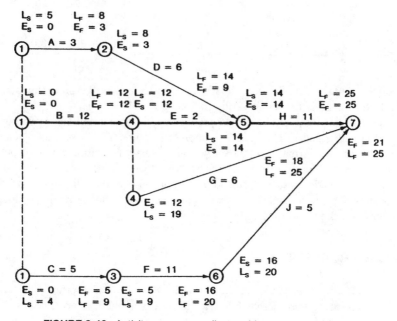

FIGURE 8-10. *Activity-on-arrow earliest and latest start and finish.*

HIGHLIGHTS

- Time estimating is challenging.
- A network diagram can provide the earliest and latest start and finish dates for each project activity, as well as the slack time for each activity.
- Schedule logically, and improve systematically.
- There are many good techniques for compressing a schedule: fast tracking, crashing, assumptions analysis, and so on.

9

Cost Estimating
and Budgeting

This chapter describes methods for estimating costs and developing the budget, starting with resource planning and cost systems.

RESOURCE PLANNING

Resource planning is the project management process of determining the types and quantities of needed project resources. Resource planning is a process that occurs before cost estimating. Typically, the project manager considers the following three questions:

- What are the requirements?
- What skills are needed?
- What is the availability of each required skill?

Notice that the ordering of the questions reinforces some principles that we have emphasized throughout this book. Customers and their requirements are always the starting point. Getting the right people on your project is a strategic issue. Getting the timing right (with regard to availability) is a constant process of balancing competing demands through analysis and negotiation. We will further elaborate on resource planning issues in Chapter 10 and in Part 3.

COST ESTIMATING

Costs may be stated in terms of the number of labor hours required, a situation not uncommon in a research group in which a certain number of labor hours have been allocated to a particular project. Cost is more commonly stated in dollars (or yen or euros), however, which entails converting labor hours into dollars. Different hourly rates typically prevail for different seniority levels, and the cost of nonlabor elements (e.g., purchases or travel) is also included. Figure 9-1 shows one way to summarize and total time-phased labor and nonlabor estimates for a task by using a spreadsheet. This figure illustrates the main elements of any successful cost-estimating system: Estimate *labor hours* (perhaps by category, for example, cost per hour, billing rate, or similar gradations) and *nonlabor dollars* for each task in each department or group involved. In Figure 9-1, the estimated entries are in larger, italicized numbers. Figure 9-2 is another spreadsheet example. There, the series of tasks comprising the project (in this illustration, from preproposal inception to postcontract closure) is listed in the left column.

Plan costs to the level of detail to which they will be reported to you.

Cost is, of course, necessary for planning a project both to sell and manage the job. In general, do not plan costs in detail greater than what you will receive in accounting cost reports. There is no point in making cost plans on a daily basis if the organization's cost reports are furnished biweekly or monthly. Cost plans, regardless of how they are arrived at, typically should be summarized in periods corresponding to expense reporting. In counting such things as travel cost and computing hours, however, work with hours or days of travel in estimating, and sort these into monthly periods.

Just as with the schedule dimension plans, there are inaccuracies inherent in cost estimates, and these must be expected and tolerated. Tolerating such inaccu-

Project: *Materials Study*   Task: *B - Obtain Materials*   Department: *Materials*

		Hours Each Month						Totals	
		1	2	3	4	5	6	Hours	Dollars
LABOR	$/Hour								
Sr. Prof.	25	*8*	*4*	*2*				14	350
Jr. Prof.	20		*40*					40	800
Sr. Tech.	15							0	0
Jr. Tech.	10							0	0

DOLLARS	Rate %								
Labor Cost									1150
Overhead	100								1150
Direct Nonlabor		*200*							200
Subtotal = Prime Costs									2500
G&A	15								375
Subtotal = Total Costs									2875
Profit	10								288
Total = Contract Price									3163

Assumptions:

Prepared By: Date:
Approved By:

FIGURE 9-1. *Typical task cost estimate using a spreadsheet form.*

Week # =	1	2	3	4	5	6	7	8	9	10	11	12	13	14	15	16	17	18	19	20
Identify prospects																				
Hours =	4																			
Nonlabor $ =																				
Meet a prospect																				
Hours =		6																		
Nonlabor $ =		50																		
Develop ideas for a project																				
Hours =			12																	
Nonlabor $ =																				
Submit proposal																				
Hours =			6																	
Nonlabor $ =			50																	
Get contract																				
Hours =			2																	
Nonlabor $ =																				
Do the project work																				
Hours =				8	16	16	12	10	8											
Nonlabor $ =					100	100	50	50												
Finish the project																				
Hours =									8	8										
Nonlabor $ =																				
Close the project																				
Hours =											4									
Nonlabor $ =																				
Evaluate the project																				
Hours =											2	2								
Nonlabor $ =																				
Follow up with client																				
Hours =													1		3		1			3
Nonlabor $ =															50					50
Cumulative Totals																				
Labor Hours =	4	10	30	38	54	70	82	92	108	116	122	124	125	125	128	128	129	129	129	132
Labor @ $50/hour =	200	500	1500	1900	2700	3500	4100	4600	5400	5800	6100	6200	6250	6250	6400	6400	6450	6450	6450	6600
Nonlabor $ =	0	50	100	100	200	300	350	400	400	400	400	400	400	400	450	450	450	450	450	500
Dollars =	200	550	1600	2000	2900	3800	4450	5000	5800	6200	6500	6600	6650	6650	6850	6850	6900	6900	6900	7100

FIGURE 9-2. Cumulative labor-hour and expense planning with a spreadsheet.

135

racies does not mean encouraging them. The goal is to be as accurate as possible and to recognize that perfection is impossible.

The more effort you put into a cost (or time) estimate, the more accurate it is likely to be. Figure 9-3 indicates the kinds of trends you might expect. In the case of capital construction projects, cost estimates range from about 0.5 to 5 percent of the project's size if you desire accuracy within plus or minus 5 percent. Conversely, the estimates require only about 0.01 to 0.10 percent of the project's size if you need plus or minus 50 percent accuracy.

Techniques

Forecasting and *estimating* frequently are used interchangeably to refer to preparing a plan for the cost dimension. Actually, the dictionary definitions of these words are different, and the differences have important implications. In project management, we are talking about the amount of money (or time) expected to be required to complete a piece of work.

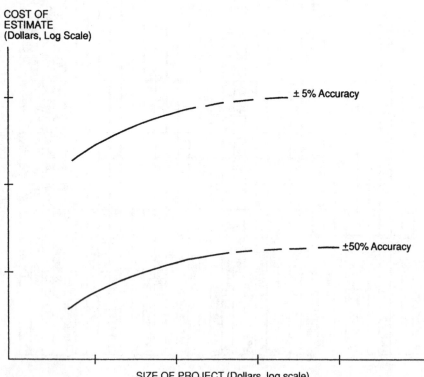

FIGURE 9-3. The expected relationship between the cost to prepare a cost estimate, project size, and the accuracy of the estimate.

If you were asked to estimate pi, you could do so as accurately as you wish because pi is a known quantity (3.14159 . . .). If, however, you were asked how long an untelevised football game will take, you probably would reply two and a half hours. You are now estimating a future event's duration based on similar previous events. You might have made this estimate by looking up the time for the longest and shortest football games ever played and by noting the times of all other football games for which durations were recorded. You would have learned that the vast majority of football games took between two and a quarter and two and three-quarters hours; therefore, two and a half hours is a reasonable estimate.

Actually, the football game you will see will not take two and a half hours. The probability of your estimate being correct is essentially nil, but you can be close. The only way to *guarantee* that actual costs do not exceed your estimate is to make the cost estimate very, very high—which probably will mean there is little likelihood of getting the project authorized. The fact that you are not going to be right means that you should become accustomed to being wrong and should not be afraid of it. But it does not imply that you should not try to be accurate. Despite these hazards, the goal in estimating is to have a meaningful plan for your project, one you can use to sell the project proposal to your customer, explain your actions to your boss, and provide enough resources to do the job successfully.

There is no point in attempting to estimate a budget for an activity until you have established its duration. **Schedule first; estimate second.** In addition, you should understand the preceding and following activities in order to define better the activity you are estimating. Such understanding may clarify that a following activity is farther downstream than it first appears. If so, the activity you are estimating probably is longer, and therefore costs more, than you first thought.

You do the estimating by breaking the project into tasks and activities using the work breakdown structure (WBS) and network diagrams. The budget of any large activity is **Estimate the cost of each task.** the sum of the smaller tasks that compose it, as shown in Figure 9-4. In general, use as much detail as possible. Every task in the WBS probably should have an individual task estimate (such as Figure 9-1) prepared by the responsible task manager.

Methods

A number of means are available to prepare cost estimates. Using as much detail as possible is commonly called the *bottom-up method,* as we discussed in Chapter 5. The major project is divided into work packages small enough to allow accurate estimation. The project estimate is the sum of the estimates for all the individual work packages.

There are shortcuts to estimating some of the small work packages. You can use similarities to and differences from other tasks to shortcut a complete level of detail for a second task. Or you can use ratios or standards to relate one small task to another.

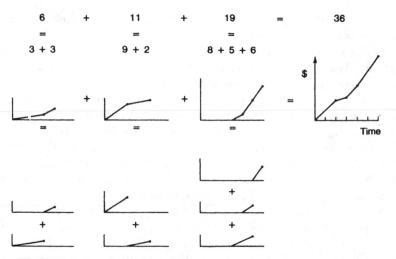

FIGURE 9-4. *A project's cost is the summation of costs for all project tasks.*

> **To make a project cost estimate, add detailed task estimates for each work department and adjust them if the overall summation seems unreasonable.**

Whenever you use the bottom-up technique, judge it against a *top-down estimate.* The top-down estimate is done first, quickly and judgmentally, and then it is set aside. For instance, assume that the bottom-up estimate comes to $10 million. Your top-down estimate, which you now retrieve, is $5 million. Go back and look at each individual work package in the bottom-up estimate to find out where the excess costs arise. Examine each package to discover to what extent there has been an incorrect assumption as to the amount of work called for. Or your top-down estimate may indicate that the total job should cost $20 million. Explore the details to find out what has been overlooked or what unwarranted simplifying assumptions were made. The role of the top-down estimate, which is obviously not accurate, is to provide a point of view from which to scrutinize the bottom-up estimate.

Parametric Cost Estimating

To introduce this technique, consider the situation in which you need to obtain some software to support your project.

Figure 9-5 illustrates a relationship between project cost and a computer program's lines of code that we might discover if we examined a large number of previous computer software projects. The shaded area in the figure would surround a cloud of points, each of which represents a particular project outcome.

If we examine these projects in more detail, we could ask how many separate modules were involved. Then the data points could be separated, and we might find a trend as in Figure 9-6, where the lines are the centroids of smaller clouds of data points.

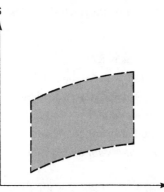

FIGURE 9-5. Relationship between project cost and lines of program code for a large number of computer software projects.

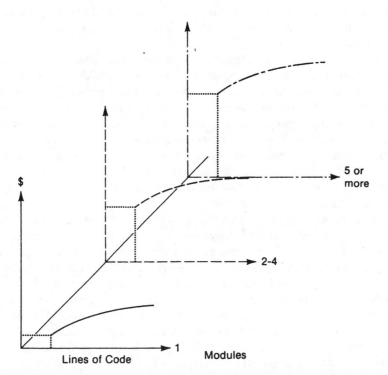

FIGURE 9-6. Trends of Figure 9-5 data when the number of program modules is considered.

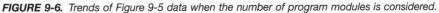

This is the underlying notion of parametric cost estimating. Historical data for many factors are used:

- *Project magnitude.* How big is it?
- *Program application.* What is its character?
- *Level of new design.* How much new work is needed?
- *Resources.* Who will do the work?
- *Utilization.* What are the hardware constraints?
- *Customer specifications and reliability requirements.* Where and how are these used?
- *Development environment.* What complicating factors exist?

Then, if your project falls within the range of these historical data, you merely insert your project's parameters into the multidimensional regression model to estimate your project's costs. There are many such parametric cost estimating models. The estimate you get from using such a parametric model is only an estimate and will be inaccurate if your input data (e.g., the resources) are poor estimates.

Cautions

As a first caution, we remind you of the Chapter 8's discussion of *padding,* the practice of expanding a task to account for unknowns. Good project managers will avoid this padding of estimates because it tends to distort the project plan and foster distrust.

Cost estimates frequently are made prematurely, before the work performance and time schedule are fully understood and defined. Such cost estimates then must be redone or adjusted when the performance and schedule are set. It is clearly more efficient to do the cost estimating after the other two dimensions are defined. However, your project definition may start with a predetermined fixed cost as the highest priority of the Triple Constraint. In that case, you must fit the best work accomplishment and schedule within that cost limitation. A good approach is to use a costed WBS, in which each task box is assigned a trial portion of the specified budget. These arbitrary initial task budgets are then adjusted iteratively as you develop planning detail.

There is a danger signal to which project managers must be alert when receiving cost estimates: the person-month dimension to describe the labor requirement. This is a danger signal because it frequently indicates a snap judgment on the part of the estimator (and not because it is an inappropriate measure, although hours seem preferable). The person who can do the task in two months may not be available when he or she is needed. Or a task that can be done by one person in three months (three person-months) may require four people if it must be completed in one month (four person-months). (So-called effort-driven scheduling is a feature in some project management software packages. If you use such software, be careful because doubling the labor on a task will automatically halve the task time.)

A pertinent issue always is how to plan for inflation (or even deflation). Such planning can be done only with great difficulty and caution. Unfortunately, there are no guarantees in dealing with the future, so some method of coping must be adopted. Make the best estimate of labor hours, regardless of when the activity will occur. Then estimate the rates for these labor hours and the dollar amounts for nonlabor ingredients in current dollars. You then can apply inflation factors for future years to these numbers by consulting with your organization's financial planners.

Another issue that you have to think about is improving productivity, which results from the proliferation of new tools such as computer-aided design and computer-aided engineering. Similar tools exist for software, many forms of white-collar office work, and so on. In many situations, these tools will shorten the time and lower the cost to carry out a given task or activity. However, before these gains can be realized, an investment is required to acquire the tools (which typically requires a capital expense) and train people to use them. Also, the learning (or experience) curve may apply, in which case future costs may be lower than current ones.

PROJECT COST SYSTEM

A project cost accounting system provides a means to accumulate costs by project and project activity or task detail. If your company does not have a project cost accounting system, you can *plan* your project costs with, for example, project management or spreadsheet software, but you will not have any practical way to *monitor* the actual costs. The following are the elements of such a system:

- Labor
  - People in your own department
  - Other people in the company
- Overhead burden
- Nonlabor
  - Purchases
  - Subcontracts
  - Travel
  - Computer charges
- General and administrative burden

There are many variations in detail for project cost systems, some of which are shown in Table 9-1. We are making the simplifying assumption that the organization has three projects (A, B, and C) that all start at the beginning and will end at the close of the fiscal year. It does not matter whether the work is for the company or for an outside client or whether it is to be paid for by the sale of goods

TABLE 9-1 Four Common Project Cost Systems, Illustrated for Three Projects ($000 omitted)

	Project A	Project B	Project C	Indirect		Total
Direct labor	50	30	10	} = 150	Labor	
Direct nonlabor	0	20	40		Purchases	
					Travel	
Overhead on direct labor				90 } = 120	Fringe benefits	} = 270
General and administrative				30	Indirect time of direct labor personnel	
Burden					Indirect labor personnel	
					Facility costs	
					General suplies	
					Publications	
Method 1						
Direct labor	50	30	10			
Direct nonlabor	0	20	40			
Direct total	50	50	50	= 150		
Burdens on direct total	40	40	40	= 120		
Total costs	90	90	90	270		
Method 2						
Direct labor	50	30	10			
Direct nonlabor	0	20	40			
Direct total	50	50	50	= 150		
Overhead on direct labor	50	30	10	= 90 }		
General and administrative				= 120		
Burdens on direct total	10	10	10	= 30 }		
Total costs	110	90	70	= 270		
Method 3						
Direct labor	50	30	10			
Overhead on direct labor	50	30	10	= 90 }		
Direct nonlabor	0	20	40	= 120		
"Prime costs"	100	80	60			
General and administrative						
Burdens on "prime" costs	12.5	10	7.5	= 30 } = 270		
Total costs	112.5	90	67.5			
Method 4						
Direct labor	50	30	10			
Overhead on direct labor	50	30	10	= 90 }		
Direct nonlabor—purchases	0	20	20	= 120		
Material handling burden	0	5	5	= 10 }		
Direct nonlabor—other	0	0	20	= 30 }		
"Prime costs"	100	85	65			
General and administrative				= 20 } = 270		
Burdens on "prime" costs	8	6.8	5.2			
Total costs	108	91.8	70.8			

or contract billing. The point is that there has to be some way of allocating the cost of these three projects to different customers or product lines. The table illustrates four methods of allocating these costs to the three projects.

In method 1, the direct labor and direct nonlabor are allocated to the project, and these are summed to provide a direct total. All the burden and overhead accounts are then lumped and apportioned to each project in proportion to the direct total expenses. In method 1, these are equal, and the billings to each of the three projects would be equal.

In method 2, the direct labor and direct nonlabor are treated as before, but the overhead portion is allocated to each project in accordance with the amount of direct labor it requires. Nevertheless, the general and administrative (G&A) expenses are allocated to the projects in accordance with the direct total, as in method 1. In this case, the billings to the projects are not equal. Project A is more than project B, which is more than project C.

In method 3, the overhead is treated as in method 2, and the direct nonlabor is treated as in both methods 1 and 2. In this case, however, all these items and direct labor are joined to come up with a prime cost, and the G&A burden is allocated in proportion to that. In this case, we arrive at a still higher amount of billing for project A.

Method 3 and its variants are the most common project cost systems.

Method 4 is one of the many common variations of method 3. Direct labor and overhead are treated as in method 3, but purchases are subject to a materials handling charge. (In method 3, this materials handling charge is included in G&A; in method 4, it is pulled out of G&A and assigned to the projects in proportion to their required purchases.) But other direct nonlabor, in this case travel, is not allocated a handling fee, as illustrated in project C. This results in a still different prime cost. Finally, the G&A expenses are again distributed, and a still different billing arrangement is arrived at.

The project manager must understand his or her company's method in order to know when to use subcontract help and when to use in-house, direct labor. It is also important to understand any subcontractor's cost accounting system. If you are placing a labor-intensive contract with a subcontractor, you should not use a subcontractor who practices method 3 as opposed to method 1.

In addition to understanding project budgets, you should understand the relationship between project budgets and administrative budgets. The *administrative* budgets include overhead, general and administrative, and capital. An o*verhead rate* typically is applied to a labor-hour rate and includes (1) allocations for the time of direct labor personnel spent on vacations, holidays, sickness, and nonproject work (e.g., meetings to deal with corporate administration, general research, and so on, (2) costs of fringe benefits, (3) allocations for departmental expenses (e.g., supplies, telephone system charges, and so on), and (4) current-year depreciation charges on *capital expenses. General and administrative* charges may be applied to labor rates alone, burdened labor rates (i.e., labor plus overhead), and purchases or other nonlabor charges. These charges consist of allocations for corporate, division, or departmental expenses that are not specifically chargable to a project.

As we saw in Table 9-1, overhead and G&A may be combined into a single burden pool. Capital expenditures enter into the overhead (or G&A) budgets by requiring inclusion of depreciation. Thus, the direct cost of a project, its own direct labor and nonlabor expenses, is not really a measure of its cost to the organization. The project must, in common with all other organizational activities, carry burdens that depend on other organizational activities and budgets.

BUDGETING COST

In prior chapters, we described the project baseline as the point of comparison for project control activities. To develop a cost baseline, the project team takes the cost estimate and distributes it to the timeline to establish its project controls. It is important to note that we have quit estimating, basically saying to ourselves, "This is the most accurate estimate that we can generate given our assumptions."

Here is an example of budgeting the cost estimate for a simple project. Let's say that there are five tasks in a project: three tasks (A, B, and C) each worth $10,000 and 10 working days, and 2 tasks (D and E) each worth $3000 and consuming 5 working days. Tasks A, B, and C are sequential and constitute the critical path, and tasks D and E (with E following D) can be completed anytime. The cost baseline is determined by when the project team chooses to schedule tasks D and E.

Let's say that the project chooses to schedule tasks D and E as soon as possible. Thus, the cost baseline (budget) for the first 10 working days is $16,000, and the budget for the next 10 working days is $10,000 and the final 10 days is $10,000. If we were to change the schedule to complete tasks D and E as late as possible, the first 10 days would have a baseline of $10,000, the second 10 days would have a budget of $10,000, and the third 10 days would have a budget of $16,000. As a final example, let's say that we want to schedule task D as soon as possible and task E as late as possible. The budget for this case would be $13,000 for the first 5-day period, $20,000 for the following 20-day period, and $13,000 for the final 5-day period.

COMPUTER SOFTWARE

One of the most attractive features of project management software is that you can obtain the overall project cost plan automatically from the task cost detail that you have entered. That is, at the same time you enter schedule information and task linkages, you may enter the resources that will be required, both labor and nonlabor. You may enter labor resources in hours, but you also must provide hourly labor rates if you wish to obtain a monetary cost estimate. Nonlabor estimates, such as purchases and subcontracts, are almost always made in monetary terms, so these will be reported in that fashion. Obviously, this is a classic garbage in, garbage out (GIGO) situation. The soundness of the resulting project cost estimate will be set by the quality of the individual task estimates.

Each software package seems to require these data in a different format, but this is not a problem if you master a single package. In general, you will have to enter all possible resources and their rates at some point if you wish to use this software to plan project costs.

TYPICAL PROBLEMS

There are several things you should watch out for in cost estimating and budgeting. First, many project groups or project managers have a tendency to make cost estimates for support-group work. This forecloses the possibility of benefiting from support-group expertise and violates the golden rule (people should be involved in estimating the tasks that they will perform). This is easily solved by requiring every department to approve the estimate for the work it will do.

A second problem is dealing with inflated estimates by support groups. Here, the project manager can first try discussion and negotiation. If that does not produce a satisfactory agreement, the project manager could alter the nature of the requested support work. Two other possible solutions are to subcontract the support work to another company or appeal to higher management.

Higher management, if it decides to "buy in," often cause a third problem. (In this usage of the jargon phrase, *buy in* means that management has decided to bid a low price for a project in order to secure the work, even at the risk of the work being unprofitable.) If you are convinced that buying in is disaster, you can request that someone else assume project management. Or you can record your objections in a memo and try to accomplish the promised work within the budget. Finally, you can undertake the job and work actively to sell your customer on changes of scope that provide an opportunity for more funding.

HIGHLIGHTS

- Resource planning is an activity that precedes cost estimating.
- Cost estimates usually are made in dollars.
- Cost estimates can be made top down or bottom up, and it is good to do both to ensure that you have a valid estimate.
- Cost budgeting is the process of developing a baseline by combining the cost estimate with the time schedule.
- The elements of a project cost accounting system, a means to tally costs by project and project task, are labor, overhead burden, nonlabor, and general and administrative burden.
- A parametric cost estimate may be useful and simple, especially if your project is reasonably similar to others for which historical data are available.

- Once all the cost estimates are summed by task, it is possible to complete a costed WBS should this be desired.
- The computer can be used to simplify planning by quickly summarizing large amounts of numerical data. For very small projects, this can be done easily by hand or with a spreadsheet and is not sufficient justification for using a computer.

10

The Impact of
Limited Resources

This chapter deals with the impact of finite resources on project plans and reinforces our earlier discussion on balancing competing demands. We discuss resource allocation and how to resolve resource constraints. Then we present techniques that allow analysis of schedule and budget tradeoffs.

RESOURCES

Projects are accomplished by the use or application of resources, namely, people and things. Human resources may include everyone in a particular organizational unit or those with a specific skill (e.g., typing, computer programming, senior optical design, analytical chemistry, or journeyman electrician). Things include any kind of equipment (e.g., lathe availability, computer time, or pilot plant time), as well as floor space to house the equipment and people. Money also may be considered a nonhuman resource. For example, a required lathe may be controlled by a model shop group or a required computer may be controlled by a data-processing group.

The project manager must organize the correct human resources to take advantage of the available physical resources. Then the project manager has to deal with the constraints and emotional problems inherent in their use while trying to accomplish the project initiator's technical performance goals within the schedule and budget.

Managing people is often the most difficult aspect of managing a project, especially for recently appointed managers whose academic training is primarily in a technical discipline such as engineering, computer science, or even construction management. Such people tend to be more

Managing projects means managing people.

comfortable with things and numbers than with people. Thus, you have to avoid the technical expert's propensity to concentrate on the quantitative aspects (e.g., engineering analyses or task budgets)—although these are not unimportant—and instead become more oriented to making things happen through people. Many good technical experts make poor project managers because they can't deal with the intangibility of people issues, such as the need to "sell" (and resell) a project to other managers. Some project managers recognize this need but cannot communicate effectively, which we discuss in Chapter 16.

The project manager must spend lots of time with people.

Resource conflicts are inherent, as illustrated in Figure 10-1. In this simple illustration, only two projects are shown. Their need for a particular resource (e.g., square feet, hours of usage of a DNA synthesizer, personnel in general, or a specific kind of human resource such as junior-level analytical chemists) is shown. A planned requirement exactly matches the available resource. However, the timing of one project is slightly altered, and as a result, the required resource no longer matches

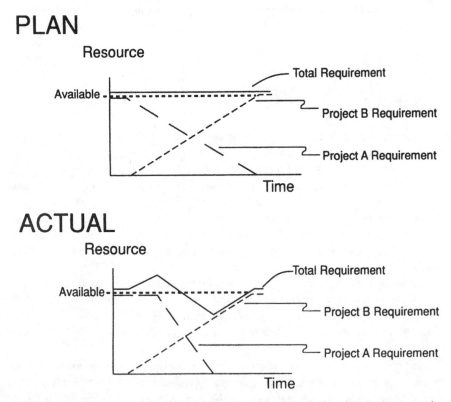

FIGURE 10-1. *Resource conflicts are inherent because the actual project resource requirements never produce a workload (on people in general or on people of a particular skill or facility) that perfectly matches availability.*

what is available. First, there is too much work, and later, there is not enough work. These mismatches are bad enough where the resources are physical things. The resulting problems often are much worse where the resource mismatch (be it overload or underload) is people. Resource overloads and underloads are a common project management issue.

Recall from Chapter 3 that resource leveling is an issue associated with balancing competing demands for resource availability and the project time schedule. Typically, the presence of limited resources causes extension in the project's duration.

Allocation

There are three reasons to consider resource allocation in a project management environment. First, forecasted use of some key resource (e.g., circuit designers) may indicate that there will be surplus personnel at some future period. This information should warn the appropriate managers either to obtain new business to use the surplus talent or to plan to reassign the involved personnel. However, many

Surplus resources waste money and talent, so resources are normally overcommitted.

resources are not very flexible and cannot be reassigned easily to other useful duties. For instance, not many highly trained specialists are flexible generalists, so an available chemist may not have useful skills for your project that requires an electrical engineer. Because no company can afford to maintain surplus resources indefinitely in anticipation of the authorization of a future project, the normal situation is that resources are overcommitted. First one resource is overloaded, then another, and so on. Sometimes the overload is only a small amount; sometimes it is substantial. Sometimes the overload is for only a short period; sometimes it is for a long period. Failure to allow contingency for this normal overload situation is one reason why many projects are delayed.

Another reason for resource allocation is to avoid inherent inconsistencies, for instance, using a particular individual on two tasks at the same time. Preparing a network diagram to a time base emphasizes resource allocation and reveals latent conflicts.

Resource capacity is limited, and multiple projects may require the same resources at the same time, resulting in a conflict that must be resolved.

A network diagram can show what resources are required and when and may reveal that more of some resources will be needed than will be available at some time. When you discover this, you must adjust the network diagram to shift the overloaded resource requirement to some other time. If you fail to do this, slippage will occur. Figure 10-2 illustrates two possible schedules for and resource allocations on a project with five tasks (A, B, C, D, and E). In this case, the resource is the personnel headcount. Tasks A and B, each of eight weeks' duration, require three and five people, respectively. Tasks C, D, and E are not on the critical path, and examination of the earliest and latest times for them shows that they can be commenced immediately or as late as the

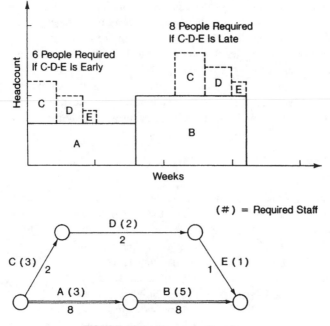

FIGURE 10-2. Resource allocation.

eleventh week. If the company performing this project employs only six people, task D would have to start early enough to be completed before the end of the eighth week, when task B is scheduled to start. If task D starts later than the start of the sixth week, some (or all) of task D will be scheduled in parallel with task B, requiring seven people.

> *Executive management constantly tries to balance the cost of idle resources versus the cost of delayed projects, and minimizing these costs may affect your project.*

A third use of this kind of analysis occurs in a large company. Imagine that tasks C, D, and E are performed by a particular support department, for instance, the design and drafting section. If the design and drafting section was provided with resource allocation information for all projects, as shown in Figure 10-2, its people could identify the earliest and latest dates at which the support, in this case tasks C, D, and E, would have to be applied. Doing the same for all projects would allow the support group to even out its workload and to identify in each case the impact of any slippage.

Removing Resource Constraints

Consider the activity-on-arrow (AOA) network diagram in Figure 10-3. After planning the work, you arrive at a summary of labor skills required for this project (Table 10-1). Suppose that you have only nine junior engineers. What are your options? What might you do? What are the risks?

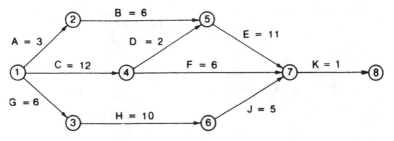

FIGURE 10-3. *Activity-on-arrow diagram.*

Figure 10-4 graphically summarizes the junior engineer staffing for each week. There are two weeks when 10 junior engineers are required, but the constraint is that only 9 are available. Some of your options are to use overtime, use senior engineers to do the work, or delay task F (which has slack) by two weeks. The latter, which seems most appropriate, runs the risk of reducing slack that may be needed later.

In general, you can do the following to remove resource constraints when only one project is involved:

1. Identify resource requirements for tasks on the critical path.
2. Add resource requirements for other tasks, using desired start dates.
3. Compare resource requirements with resource availability.
4. Identify options to remove resource conflicts that are found. You may verify that the conflict is real, adjust start dates for tasks with slack, give the schedule more attention to reduce downtime inefficiency, improve productivity (with new tools, improved match of people with tasks, or incentives), adjust resource

A project schedule that requires use of already assigned resources is unrealistic.

TABLE 10-1   Resource Allocation Using the Network of Figure 10-3

Task	Planned Duration	Senior Eng'r	Junior Eng'r	Design	Mech	Elec	Inspect
A	3	2	4	4			
B	6		2			3	
C	12*	3	4	4			
D	2*		2		5		
E	11*	2	5		2	2	
F	6		2	7			
G	6	4	1				
H	10	1	3				
J	5		2				
K	1*	2	2				3

*Task is on critical path.

Task	Junior Engineer Required Each Week (Assuming Earliest Start on Each Task)
A	4 4 4
B	2 2 2 2 2 2
C	4 4 4 4 4 4 4 4 4 4 4 4
D	2 2
E	5 5 5 5 5 5 5 5 5 5
F	2 2 2 2 2 2
G	1 1 1 1 1 1
H	3 3 3 3 3 3 3 3 3 3
J	2 2 2 2 2
K	2

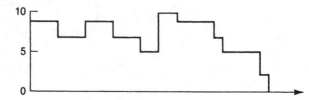

FIGURE 10-4. *Junior engineer resource requirement for the project in the network diagram of Figure 10-3.*

availability temporarily (by planning overtime, rescheduling vacations, obtaining temporary employees, or subcontracting tasks), change resource availability (by training or promoting surplus skill to fill needs or hiring new people), change the time schedule (accelerate the critical path, which will put you ahead of schedule, delay the critical path, which will make you late, or change the start date of new projects), or change the plan (specifications, task sequence, or standards).

Suppose that you have a simultaneous second project (Figure 10-5). Suppose that task W requires four inspectors, and you have only six. That's no problem on the second project taken alone, but it is when you try to accommodate its need for inspectors with the first project's need, as shown in Table 10-1. Both task W in the second project and task K in the first project are on the critical paths, and they need a total of seven inspectors at the same time. Something has to be done because you have only six inspectors.

The procedure in this case is the same as removing conflicts on a single project, except that

1. Projects must be ranked by priority.
2. The highest-priority project gets the first claim on available resources.
3. The second-priority project gets second claim, and so on.

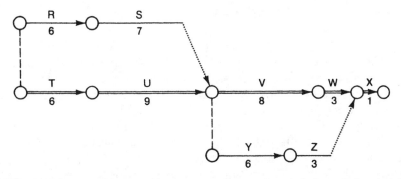

FIGURE 10-5. *Second project requiring resources that are also needed on a first project. (This project starts at the same time as the project shown in Figure 10-4.)*

The result is that lower-priority projects usually get delayed, performance is compromised, or there is a lot of (unplanned) subcontracting. Sometimes it is cost-effective to accelerate a small or low-priority project to get it out of the way and thus avoid a conflict that otherwise would arise with a major or more important project.

Lower-priority projects usually lose the competition for limited resources.

Project schedules usually are prepared initially without regard to whether the required resources actually will be available when desired or required. Thus, there can be a serious problem if the impact of resource constraints is overlooked. The first step to avoid this problem is to refine the schedule for your project so that all tasks are consistent with available resources. Then the resource requirements of other projects must be checked and conflicts resolved. These other projects include both those that might start and be in process during your project and those existing (or planned) projects that are supposed to be completed before your project starts but are delayed and thus have an impact on your project. If this is not done, the lack of resources will not cure itself magically; it will become an obstacle when there is less (or no) time to devise an alternative schedule.

COMPUTER SOFTWARE

Most project management computer software will produce resource histograms similar to the lower part of Figure 10-4. Software packages also will level resources or resolve resource overloads automatically. However, the software algorithms for leveling do not necessarily produce a practical schedule. Therefore, this resource-leveling capability should be tested and evaluated carefully before you apply it on your project.

A network diagram is not merely a schedule dimension plan; it also clarifies resource allocation.

TIME-VERSUS-COST TRADEOFF

The critical path method (CPM) historically has been associated with network diagrams in which there is considered to be a controllable time for each activity. This implies that activities can be accelerated by devoting more resources to them. Thus, there is a time-versus-cost tradeoff for each activity and consequently for a path or the entire project. Figure 10-6 shows this kind of situation.

If you are trying to accelerate a project, you should focus on the critical path. Of all the activities on the critical path, the most economical to shorten are those with the lowest cost per amount of time gained.

Consider the following situation that you might face as the project manager for the project illustrated in Figure 8-10.

To carry out this project (assuming that all times are in weeks), you must rent a standby electrical generator for the entire duration of the project (however long it is) at a cost of $1,000 per week. Your purchasing department has told you that the subcontractor performing task B has offered to shorten its performance time by as much as 5 weeks (i.e., to 7 weeks) but will charge $800 per week for every week less than the original 12 weeks (i.e., a premium charge of $800 for 11 weeks' delivery, $1,600 for 10 weeks' delivery, or $4,000 for 7 weeks' delivery). You can save $200 per week by accepting the subcontractor's offer. However, you have to look at your network diagram before rushing to accept the offer (Figure 10-7).

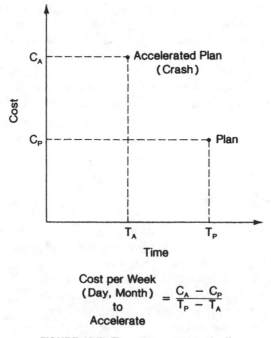

$$\text{Cost per Week (Day, Month) to Accelerate} = \frac{C_A - C_P}{T_P - T_A}$$

FIGURE 10-6. Time-versus-cost tradeoff.

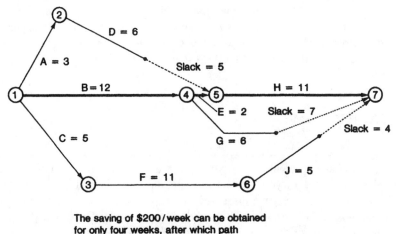

The saving of $200/week can be obtained
for only four weeks, after which path
C-F-J also becomes a critical path.

FIGURE 10-7. *Time-based activity-on-arrow diagram showing that the savings of $200 per week can be obtained for only four weeks, after which path C-F-J also becomes a critical path. (Depending on the certainty you feel for C-F-J and B-E-H, it might be better to shorten activity B by only three weeks, thus maintaining only one critical path.)*

As you can see, once task B is shortened from 12 to 8 weeks (at a cost premium of $3,200, producing a $4,000 saving on the standby generator), there is a second critical path. Thus, you cannot advantageously shorten task B by 5 weeks, only by 4. In fact, you might prefer to shorten task B by only 3 weeks to avoid having two critical paths.

Figure 10-8 shows another aspect of this. The direct cost curve depicts those costs associated with carrying out the project that are time-dependent and for which

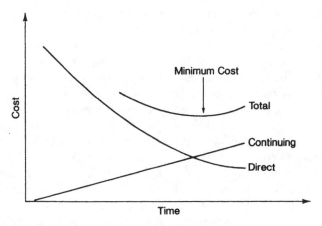

FIGURE 10-8. *Finding the lowest cost.*

there is a cost premium associated with shortening the program. In addition, there might very well be continuing costs associated with the program, for instance, the rental of standby power generators or other such facilities. In this kind of situation, there will be a time that leads to the lowest cost for the project.

TYPICAL PROBLEMS

A major resource problem occurs when project schedules change. This may be caused by a change in your project or in another project or by a shift in the start date of a project. As we discussed in Chapter 3, you can never be certain which proposals will be approved or when they will start. Therefore, you must be constantly alert for these changes and their impact on the resources you plan to use.

When first starting a project, the project manager and his or her team will devise a project plan that details the tasks and milestones along the path to completion. Time estimates then can be placed on each activity to determine the project completion date. These estimates are based on the assumption that the resources—primarily people in the organization—will be available both at the beginning of and throughout the specified tasks. However, as time and the project progress, priorities shift within the organization. Here is an example cited by one project manager:

> You are managing a project for which you have been promised specific, experienced personnel to perform certain critical tasks. The project has very demanding schedule and cost constraints. When the time comes to perform these tasks, the promised personnel are no longer available. You are told that you must use one of the (less experienced) personnel who are not busy.

This kind of situation arises because promised resources usually are moved on to higher-priority projects (which often means those that are in bigger trouble). This then results in a change in the project's completion date. Project managers have to anticipate and plan for such resource deficiencies. Otherwise, they often will be in the uncomfortable position of trying to justify a schedule slip to the sponsor when it was not due to any sponsor action.

Another problem can occur when you wish to use help from some other function within your organization. Imagine that your project needs help from functions A, B, and C (whatever these departments or divisions may be). The current workload forecasts for those functions are shown in Figure 10-9. The time periods might be days, weeks, months, quarters, or years. If the work you have for each other function demands exactly the amount of uncommitted time shown for function A, you should anticipate the following reactions: Function A will be able to help, and your requirements will exactly (and helpfully) fill

(continued)

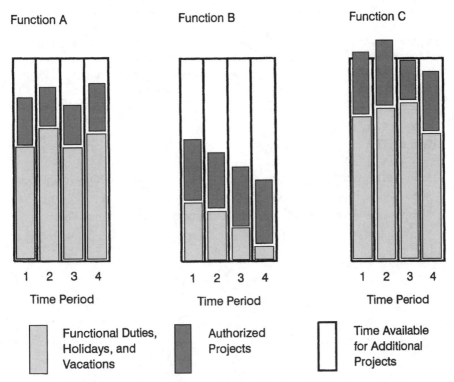

FIGURE 10-9. *Workload forecasts for three functions for four time periods.*

up their unused time. Function B is likely to *really* "help" you, saying that you don't truly understand the task's difficulty. They will try to persuade you that the task requires far more work than you thought, and they are likely to try to inflate your needs to fill up all their unused time. Function C will be unwilling even to talk with you because they already have too much work. Therefore, understanding the workload in other functions can be important if you need their assistance.

There are also several subtle problems for which you should be alert. Finding highly qualified team members is always a problem because there are never enough. You often will have to use what's available rather than what's ideal. Breaking any resources loose from their last project always takes more time than anticipated. Getting appropriate time from individuals who have only a part-time assignment to your project can be a challenge. When confronted with this, try to schedule full days rather than parts of days.

Computers are very helpful for manipulating and summing personnel resource requirements on many simultaneous projects and thus can help to alert you to prospective resource constraint problems. When computers are used to

(continued)

assist with planning (or anything, for that matter), there is always the danger of entering incorrect data or making programming errors. With standard project management software, the software is proven, but there still can be incorrect data entry. To avoid this kind of problem, verify data entry or run manual spot checks of output. Never blindly accept any computer output as gospel.

HIGHLIGHTS

- Resources, whether people or things, should be allocated carefully in a project.
- A resource histogram can clarify resource allocation.
- Each activity, critical path, and project has a time-versus-cost tradeoff.
- Computers can aid project planning in several ways.

11

Project Risk and Issues Management

Managing risks and issues is fundamental to successful project management. This chapter will help you learn how to engage the project team to identify risks and issues, analyze them, develop responses, and implement those responses.

One of the differences between excellent project teams and mediocre ones is this: Mediocre project teams plan only for success, whereas excellent project teams recognize that there are many threats that will undermine performance. In recent years, the project management profession has become increasingly aware of the impact of failed projects and has expanded its repertoire of practices and tools for project risk management. You will increase your chance of having a successful project if you integrate risk and issues management into project planning and execution.

One way to consider project management maturity is the organization's awareness that "failure is a possibility."

We will begin our discussion of this vital topic with the professional definitions of risks and of issues.

A *risk event* is a *discrete occurrence* and is describable as a cause and its associated risk consequence. There is a probability that the risk event will occur. A useful criterion for recognizing a risk is: "Maybe it will happen, or maybe it won't." *If* that risk event occurs, *then* there will be a consequence of that risk. All three elements (event, probability, and consequence) are necessary to define a risk.

In contrast to the discrete and *probabilistic* definition of risk, *issues* are *certainties* and are evaluated and managed differently than risks. If something has a 100 percent probability of occurrence, it is not a risk; it is a certainty. Therefore, if you're sure that it is going to happen, or if it has happened already, it is an issue, and you must resolve the issue proactively.

For an example of risk, consider the following: It is the start of your day, and it is partly cloudy. You check the weather forecast and see that there is a 50 percent chance of thunderstorms today. You define the risk event as "thunderstorm" and assess the probability of occurrence and the potential consequences. Prudently, you decide to include a contingency provision in the day's plan: You take an umbrella with you in case the risk event occurs.

For an example of an issue, assume that you look out the window and see that it is raining heavily. Rainfall is no longer a risk but a certainty that you must manage. As a certainty, the issue demands resolution by accepting ownership and actively working the issue. As a rational and prudent person, you do not foolishly deny the presence of rain or think that you can superhumanly dodge the raindrops. You put on your raincoat and open your umbrella!

TEN STEPS FOR TEAM-BASED RISK MANAGEMENT

Risk management is best done collaboratively.

All good project teams have a collaborative process for risk identification, analysis, and response. Figuratively, these projects have "radar" that looks out over the horizon for incoming threats. The radar allows the project to anticipate threats, make decisions to prioritize the threats, and apply appropriate countermeasures to effectively avoid or mitigate the risk event. The teams that can anticipate risks and respond effectively are more likely to succeed.

Step 1: Prepare

Perform this step before assembling the team in a risk management meeting. Good performance depends on cross-functional contributions; thus, it is essential to perform risk identification collaboratively and systematically in a focused session, *not* with the project manager shuttling from individual to individual in one-on-one meetings.

A certain amount of homework is necessary before any team meeting. Here are four recommended tasks:

- *Review some fundamental questions.* Do you have a real project? (Many projects are launched without management authorization in the form of a project charter, with little to no requirements capture work.) Do you have access to the right people? (Many people do not participate in the early stages of project planning because their experience is that their assignment will change; thus, they are reluctant to participate in anything that is not product design or product launch.) Have you invited key suppliers? (In our experience, our biggest frustrations occurred when we did not get suppliers involved in projects early enough.) Do you have cross-functional representation? (A common mistake is delaying the involvement of the "downstream" functions such as procurement, testing, sales, and manufacturing.) How realistic is the value proposition of the proposed product? (The quickest way to create cynicism is to champion an idea that does not make sense for an existing or potential customer. If you are creating a radical innovation or trying to disrupt a market, make sure that people understand that purpose.) How is success defined and measured?

- *Identify stakeholders.* There are always multiple project stakeholders. Examples of commonly overlooked stakeholders include product service and support, legal, purchasing, warehousing and distribution, advertising, logistics, safety and environmental compliance functions, insurance, and sometimes important subcontractors. *Each has a different perspective on risk and a different risk tolerance.* Many stakeholders are not considered initially, and their eventual outspokenness sparks many problems. Discipline yourself to examine a wide range of stakeholders' perspectives; you will be rewarded with powerful insights.

- *Research past projects.* Organizations tend to make the same mistakes. The goal is to avoid repeating past mistakes so that the project team can focus on innovating. From past project learnings, there should be a body of knowledge of common mistakes and problems, which typically include unstable or changing requirements, resource flux, unclear roles and responsibilities, and problems with interfaces. The preceding items are frequent frustrations, and you need to establish a vision that "this project will be different." You also should try to identify opportunities to leverage other organizational capabilities and to create new technologies, strategies, and intellectual property.

> *No project completely goes in accordance with the original plan. What you don't know when you start is how and when your project will depart from plan.*

- *Communicate the purpose.* You need to ensure that the participants understand the purpose of the meeting. Send out your announcements well in advance. Allocate sufficient time for the session; it is easier to end a meeting early than it is to schedule the additional time needed to finish risk-response planning. While you will hear, "I don't have the time," remember that the adverse impact from one risk event easily can cause weeks of schedule delay and overrun. Recognize that there is a learning curve: If people are new to the risk management process, plan for *at least* a day. For larger and more complex projects, it may even take longer. The team will become more efficient as it develops fluency with the terms and tools. These decisions happen best in face-to-face environments.

In today's fast-paced global environment, people often meet virtually. While we have seen some successes, most virtual meetings are poorly facilitated. Meet face to face at one location until you have built an experienced project cadre. Risk management is fundamentally a process of communications and group decision making.

Step 2: Build Communications with Common Language

Common language creates grounding for good communication and prompt action. You should review and "level set" the group on the definitions of risk and issues found in the beginning of this chapter, as well as the risk-response strategies of avoidance, mitigation, transfer, and acceptance as we will describe in step 7.

Start with a risk-identification icebreaker, asking the assembled team this question, "Consider our firm, as well as others. What makes projects perform poorly (or even fail)?" The answers foster a common mental model about the strategic aspects of the project. In addition, the question stimulates thoughtfulness and deeper probing into the complexities of the project. Record the output on flipcharts, and hang up the charts to remind the team of sources of potential concern. This icebreaker should only take 15 minutes.

Ask executives to participate in the icebreaker exercise because it "decriminalizes" risk, helps to point out the differing paradigms that people apply to performance, and sets the stage for effective risk allocation. The goal is to develop leverage: The team needs to impress on the executive what the potential risks and issues are, that *some* of those risks and issues require executive ownership, and that the executive needs to support the decision making process by indicating priorities.

This icebreaker is a top-down approach to thinking about failure. It widens the perspective of subject matter experts. Most people, especially those with strong technical backgrounds, tend to approach projects from a narrow specialist perspective and be unconcerned with the integrated functioning of the system. By directly addressing success and failure, you can get them to focus on the strategic issues and not insular details.

> **Be realistic when making decisions and optimistic when implementing them.**

Good leaders establish an effective culture for the team, and risk management can be a catalyst for that. An effective team culture has many characteristics, and a few of these characteristics include open mindedness, honesty, authenticity, balance of inquiry and advocacy, patience and consideration for others, and passion and willingness to commit. Paradoxically, organizational culture in many U.S. organizations (e.g., be optimistic, confident, cheerful, competent, and cooperative) is a source of risks because many people are wearing a "happy face" mask and suppressing their reservations. Leaders have to help individuals overcome their reluctance to raise concerns that stem from their fear that others will label them as "not being a team player" or "not being positive." A supportive team culture helps to ensure that people focus on the success factors rather than on developing defensive routines that result in finger pointing.

The biggest dangers result from *errors of omission* rather than *errors of commission.* An example error of omission is lacking or ignoring essential technical or functional input, whereas an error of commission is allowing a design review meeting to turn into a problem-solving meeting.

Case-study analysis, personal experience, and review of empirical studies suggest that there are a number of predictable, recognizable red flags that signal poor project performance. You might want to use the simple diagnostic instrument illustrated in Figure 11-1 to help you to identify and list a number of common causes of frustrations. Assign the indicated weights to each "True" item; if the total is

Red Flag	True	False	Weighting
Insufficient staff			10
Lack of market information			9
Dysfunctional teamwork			9
Insufficient financing			9
Unrealistic time frame			7
Lack of focus			6
Lack of management support			5
Technical limitations			4

FIGURE 11-1. Red flags for projects. A score greater than 25 indicates a project heading for trouble.

greater than 25, you probably should change the project plan or organization. You can use this tool to assess your project's likelihood of getting into trouble. Feel free to change the weighting and the factors to suit your own situation. Look deeper for the root causes of the frustrations.

Step 3: Generate a List of the Team's "Concerns"

The goal of step 3 is simple: Anticipate threats to success by generating a list of risk events and a list of issues. By recognizing hazards, the team can take steps to avoid them, react more quickly and effectively, or turn the threats into opportunities.

Figure 11-2 presents a list of proven risk identification techniques. These tools help individuals and teams to reframe their mental models and avoid potential decision-making blind spots. Note that there are common disadvantages to each of the techniques that include they take time to learn and time to perform and create psychological discomfort that may result in interpersonal conflict. Some of these techniques are common, and some are not.

As the team generates concerns, it should segregate the concerns into risks and issues. Risk events are discrete and meet the criterion of "Maybe it will happen, and maybe it won't." Issues are statements, sometimes vague, of concerns that "have happened," "are happening," or "are certain to happen."

Many people often regard the difference between risks and issues as an unimportant fine point. Here is a typical reaction: "At first, team members don't see the relevance of separating risks from issues. No matter how many examples you give

Technique &Description	Advantages	Disadvantages
Assumptions analysis – Assumptions are factors that, for planning purposes, are true, real, or certain. These assumptions may be invalid during implementation. Assumptions are incrementally altered, or totally invalidated, and the resulting impact is assessed.	Can support with quantitative analysis (e.g., sensitivity analysis)	Takes time to identify, discuss, and understand each person's "mental models"
Brainstorming - The cross functional team generate ideas about project risk, often done with the support of a facilitator	Energizing and creative Easy, familiar	Results may be perceived as ambiguous and difficult to organize
Checklisting - Based on historical information and knowledge of the system. Helps to make sure that past mistakes are not repeated.	Quick.	Tedious, often overwhelming, and may result in uncritical "checklist mentality. Not prospective
Delphi - This technique is best used for complex messy problems where there are numerous perspectives on the nature of the problem and the solution. For example, a question could be "what is the future for electric powered vehicles."	Remotely extracts expert opinions and builds consensus	Time consuming Results are often ambiguous
Document reviews - Assemble and analyze documents for content (literal meaning) and context (application) Initial plans, assumptions, past experience, statements of strategy, requirements specifications and review.	Establishes common basis for further analysis and decisions	Tedious
Diagramming - A variety of techniques to make models of stock, flows, causality, decisions, and assumptions more explicit. (Diagramming techniques include fishbone, flow charts, influence diagrams, relationship mapping).	Analytical Complements other techniques	Difficult to get agreement on accuracy.
Independent assessment - External assessors use a variety of techniques to inspect the project	More objective, comprehensive Domain expertise often required	Less ownership
Interviewing - An analyst elicits knowledge from others through guided questions and probes	Depth of inquiry and follow up questions possible	Bias
Triggers - A list of symptoms or warning signs that indicate a risk event has occurred or is likely to occur. Analogous to dummy lights on your automobile's dash board	Effective and efficient once implemented	Requires integrated understanding of system and tolerances

FIGURE 11-2. Identification techniques.

them to define risks and issues (e.g., a flat tire is a risk, and driving on a near-empty gas tank is an issue), people will still struggle with distinguishing between risks and issues." However, experience shows that the distinction is very helpful.

Try to make the concerns as specific as possible, and you will save the team a lot of frustration in the later steps. The cause-effect structure, as illustrated in Figure 11-3, is helpful in understanding and recognizing a risk. This example shows the outcome or effect called "Delayed Product Launch." Note the two potential causes, which are typical of what emerges during risk-identification brainstorming. They are risk events because they meet the test of "Maybe they will happen, and maybe they won't." The "Supplier Late" factor results from deeper potential causes, and

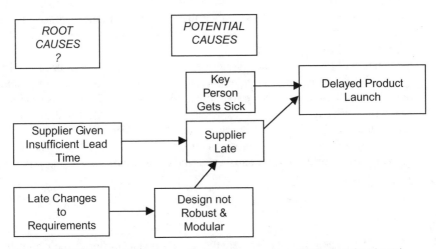

FIGURE 11-3. *Example of a cause-effect diagram for a delayed product launch.*

this is where the analysis gets interesting. The item entitled "Supplier Given Insufficient Lead Time" is not a risk; it is an issue that needs to be managed through the issues management process. Thus, by not providing sufficient lead time, the organization has created the potential for a delayed launch. Now, the team can step up and work, in a partnership, to ensure that the supplier can meet his or her obligations. We could make the similar argument about design: If the supplier does not have a modular and robust design capability, then it can't respond to late changes in requirements. To minimize risk, the team needs to look across the entire supply chain and work proactively to build capability instead of finger pointing and blaming suppliers.

Diagramming techniques can be helpful for modeling the chain of cause and effect. In addition, diagramming techniques help to point out the systematic nature of product development and emphasize that responsibility for success is shared by many stakeholders.

A common question is: "When do I stop the cause-effect decomposition?" The answer is to work backward through the chain of causes to a point where an individual has influence on the risk event's occurrence. When the team has judged that it has analyzed to a sufficient level of granularity, stop and document the risk event. A list of discrete, specific risk events is the input to the quantification and assessment process.

Generally, early in the project you will find a larger quantity of issues than risks. This is so because the team usually has an incomplete understanding of the product development effort, so it is hard to develop specific cause-effect statements, and natural language is ambiguous. Nevertheless, the real value of the process is getting the

Analysis aids understanding and improves project outcomes.

cross-functional team members to articulate their concerns in language acceptable to everyone else. With practice, the team will develop the skill to extract specific risk events from issues.

This ideation of concerns—and stating them specifically—does not have to take a long time and often can be done in less than an hour. Often, the team is impatient and reluctant to do a comprehensive identification of risks. However, it is possible to sustain energy by reminding the team that if a risk is not identified, the team ends up with a workaround that will consume precious time, effort, and money.

You now have a list of team-identified concerns for further analysis. The more specific you can make these risks and issues, the better it will be for analysis and response planning.

Step 4: Classify

In step 4, the team takes its list of risks and list of issues and evaluates their sources. As the team works through the analysis, the risk-issue partitioning helps the team to see that it *collectively owns* the responsibility for the success of the project. The risks and issues belong to the team, and the team can't rely on wishful thinking. The process of listing risks and issues helps the project manager build an effective project plan that has buy-in. Often, the team can resolve issues simply by adding a task to their work breakdown structure (WBS) and schedule.

The classification process helps the team to clarify what it knows and what it doesn't know, so team members can deepen their analysis and determine an action plan. We find that the following labels work very well for classifying the sources of project risk:

- *Technical risk sources.* These are sources associated with the design or operation of the project's product and/or production processes.
- *Logistic risk sources.* These are sources associated with regulatory or economic changes, supply, procurement, inventory, maintenance, and support.
- *Programmatic risk sources.* These are sources associated with obtaining and applying program and project resources such as technical experts, project specific tools, and budgets for capital and project expenses.
- *Commercial risk sources.* These are sources of risk that change assumptions and affect revenues, costs, market share, profitability, and so forth.

Avoid classifying risks by functional area (e.g., engineering risks, manufacturing risks, and marketing risks) because this practice perpetuates finger pointing and blaming others. Most project and product performance problems occur at interfaces.

Often, it is worthwhile to deepen the analysis to determine "strategic" and "operational" factors. *Operational* risks are those that affect delivery time and development expenses. Technical problems usually are operational, not strategic. *Most* technical problems *can* be solved given sufficient time and resources. The team is developing products with time-functionality-cost tradeoffs. While in technical development, a team will spend most of its time with operational risks.

However, do not overlook the *strategic* concerns that affect project selection or the ability of the project to meet the business objectives, such as profitability or customer satisfaction. Periodically, the team should evaluate risk events against the customer-value proposition. Project cancellation is a legitimate response to some risk events or issues.

Step 5: Analyze the Risk

Recall that risk events are discrete events that meet the test "Maybe it will happen, and maybe it won't." Risk analysis is the process where the team estimates both the probability and consequences of a risk event. The word *impact* is frequently used synonymously with the word *consequence* when discussing risk. The output of the calculation is an estimate of the overall risk exposure. The formula for a risk exposure is probability of the event multiplied by consequences of the event equals risk exposure, which we abbreviate as

$$Pr_{event} \times consequences_{event} = risk\ exposure$$

Figure 11-4 illustrates a *risk map* (also called a *probability-impact matrix*) showing six risk events arranged by probability and impact for overall risk. There is one "High" risk, four "Medium" risks, and two "Low" risks.

Figure 11-5 shows the format of a table you may wish to use for structuring an analysis of risk.

The quantification process helps the team to deepen its understanding about which risk events require more attention. Recall the words of Lord Kelvin, "When you measure what you are speaking about and express it in numbers, you know something about it, but when you cannot express it in numbers, your knowledge about it is of a meagre and unsatisfactory kind." Teams can scale the rating by anchoring the extremes to project success and failure definitions developed during the icebreakers in step 2. Assessing the risk impact helps the project team to evaluate critical success factors.

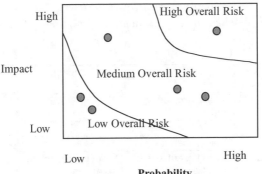

FIGURE 11-4. *A probability-impact matrix for evaluating overall risk.*

Event	Pr_{event}	Consequences_{event}	Risk Exposure (Pr x Consequences)	Rank

FIGURE 11-5. *Chart for structuring risk analysis.*

Some individuals will object to the large number of potential and trivial risk events that may be present. They fear that the volume may overwhelm them with work and waste their time. While this may be true, remember that you need to develop and apply judgment as to what is strategic and what is trivial with respect to the success of the project.

We offer this caution when you see very high probabilities: You may be looking at an issue rather than a risk, and you will want to use the issues management technique rather than the risk management technique described in step 7. In our experience, there are few technical risks in projects; rather, there are product performance issues that are resolved through design tradeoffs.

Recall that operational risks are those that affect development time and expenses. Figure 11-6 is a table for assessing project operational risks in each of four potential impact areas: development expenses, target product cost, schedule, and product performance. It is important for team members to anchor impact ratings to their own business environment and scale them *specific to each project and its organizational context* (e.g., a two-week delay may be unnoticed in some companies or projects and disastrous in others).

Understand your stakeholders' risk tolerances.

Radical innovations need larger risk tolerances, whereas incremental innovations typically have smaller tolerances. Incremental products have more known and knowable items and thus more predictability. For example, stakeholders for a radical innovation might tolerate a 50 percent slip in schedule or budget, whereas stakeholders might consider a 10 percent slip as high impact for an incremental innovation.

Figure 11-7 lists several project risk-quantification techniques that the project team can use to understand the probability and impact at a deeper level.

Project teams often perform *structured reviews* (also called *red teams* or *walkthroughs*) to enhance their risk analyses. The team invites an independent

Impact / Impact Area	Low	Medium	High
Development Expenses	<15% overrun of target	15%-30% overrun	>30% overrun
Target Product Cost	Insignificant overrun	3-5%	>5%
Schedule	Insignificant delay	> 6 weeks	>3 months
Product performance	Insignificant loss of functioning	Defects present but manageable	Significant performance issues

FIGURE 11-6. *Impact table for assessing operational risks.*

Technique Description	Example Application in Projects
Decision analysis/decision trees - Applicable to selected discrete events selected and limited risk outcomes	Investment in well defined alternatives
Expected Monetary Value (EMV) – When combined with decision analysis can assess consequence of risk (Pr x Impact = EMV)	Selection of product configuration options
Failure Modes and Effects Analysis – Starting with a functional analysis of a product or process, the team identifies failure modes, assesses severity of the failure mode on the customer, assess probability, and assesses detection.	Used to evaluate the reliability of products and processes; common in complex mechanical products
Monte Carlo – A computerized simulation of project model for assessing probability of meeting time and cost targets	Develop estimates and confidence intervals for completion dates and expenses.
Program Evaluation Review Technique (PERT) - Useful project technique for probabilistic evaluation of time and cost. It requires some estimate of variance in the input data.	Estimate confidence in time and expenses by estimating an average project duration and a variance. The variance is a quantitative measure of cost or schedule risk.
Real Options Analysis– This technique allows managers to identify risk components and decide which ones to hold, hedge, or transfer. It preserves flexibility and allows for frequent adjustment. It can lead to different conclusions than those from traditional discounted cash flows.	Investment decisions as projects progress through development funnel.
Scenario analysis – Develop different alternative states of the future	Strategic planning of different levels of competitive response
Sensitivity analysis – Examining the degree of change of an output with the change in inputs. The more sensitive measurements are perceived as more risky.	Common for testing assumptions in business case for pricing and profitability analysis. Easily done with a spreadsheet.

FIGURE 11-7. *Risk-quantification techniques.*

reviewer to probe the validity of estimates, boundaries, and interfaces; confirm the categorization; specificity of risks; and confirm ownership. The team retains the responsibility for project performance but gets a more objective external perspective on its true status. The review forces the project team to reflect, develop priorities, and avoid rushing to premature closure.

Chapter 21 describes decision trees, which often are combined with risk analysis as a technique to determine a rational approach to a choice that will lead to uncertain outcomes.

A final and important point is to beware of analysis paralysis! It is common for people to get lost in details and quibble over small, subjective numbers. You should watch for the team bogging down in trivia. The goal is to developing a common understanding of risks and issues.

The purpose of analysis is insight for improvement.

Step 6: Prioritize Risks and Issues

A typical project easily can have hundreds of risks and issues, which is more than any team can or should try to manage. The next step is for the team to prioritize the classified risks and issues. There are a number of techniques available for prioritization. Weighting is probably the most common technique.

It is essential to determine stakeholder risk tolerances as part of the prioritization process. It is vital for the project team to understand *stakeholder risk tolerances.* As a decision-reaching process, the project team discusses the results and establishes a "cutoff level" to determine which risks the projects will accept and which the project will try to avoid.

While it might seem logical to determine stakeholder risk tolerances *before* quantification and prioritization, we find that teams usually wait and develop the analysis and then discuss the data with stakeholders. In looking at the range of potential outcomes under various scenarios, the stakeholders look directly at the relevant data and make an intuitive decision on what they can and cannot accept. Here are some good questions for assessing the individual or organizational risk tolerances:

- What is the magnitude of the upside and the downside? When do the risks occur?
- Can we reverse or recover losses? How much control do we have?

Step 7: Plan Risk Responses and Manage Issues

Risk-response planning (step 7) is the process of developing strategies for reducing threats to the intended project and product outcomes. The team takes their prioritized risk list and begins response planning for the top-ranked risk.

Frequent fliers know that most commercial airlines insert schedule buffers so that flights usually arrive on time. We have experienced occasions when the plane departs the gate 30 minutes later than posted and still arrives at the scheduled time. In addition to schedule buffering, pilots use response options such as flying faster that allow them to meet their arrival times. Airlines manage customer expectations by setting conservative, buffered schedules and then empower their pilots (who are analogous to a project manager) with resources and authority to develop and apply responses.

Before delving into the practice of risk response planning, we want to discuss the too-common approach of *passive acceptance* of risks, which is defined as accepting the consequences of a risk by ignoring it or practicing wishful thinking. Consider this comment:

> People tend to do the easy stuff on their projects first, and then they sweep problems under the rug. A good leader will focus first on the tough things. When teams "sweep problems under the rug," they have passively accepted the risk event.

The effective project manager leads and encourages participation in initiating and sustaining the effort to manage project risks. This includes recognizing that organizational culture often allows individuals to stay in their comfort zone, which is often functionally siloed. This comfort zone often creates psychological denial and weak self-responsibility, as well as excuses for firefighting and crisis management.

There are four generic risk strategies that the team should consider (listed below). Also, note the questions that can help you to evaluate the merits of each of the strategies.

- *Risk avoidance.* This involves changing the project plan to eliminate the specific risk events or conditions. By avoiding the risk, the project team removes a source of poor product or project performance. Avoidance also can include project cancellation. However, by practicing avoidance, the organization may miss opportunity. You might ask the following questions to apply risk avoidance: How could we avoid this risk event? Are the hazards associated with this risk event so significant as to justify canceling the project?
- *Risk mitigation.* This reduces the probability or impact to an acceptable threshold. Strengthening a product's reliability (e.g., reducing mean time between failures) is an example of reducing the probability, and robust design (building in redundant or backup systems) is an example of reducing the impact. You might ask the following questions to apply risk mitigation: How could we reduce the probability of the risk event happening? How could we reduce the impact of the risk event on the project's success?
- *Risk transference.* This involves moving the responsibility for the risk to another party. Examples include joint ventures, subcontracting, and purchasing insurance and warranties. Transfer is often the most "out of the box" concept for many teams, who traditionally have thought of the work as one of technical and market development. You might ask the following questions to apply risk transference: Could another organization handle the risk better? How could we contract out the risky work or the risk event?
- *Risk acceptance.* Active acceptance is to develop a contingency plan. Passive acceptance leaves the project team to work around risks as they occur. You might ask the following question to apply risk acceptance: If we have to accept the risk, what contingency strategies are available to us?

In Chapters 8 and 9, we presented some of the reasons that projects should not pad individual tasks in their estimates in order to protect against threats. Instead, good practice is to estimate tasks accurately (the PERT method described in Chapter 8 is a good way to do so) and then compute a "buffer" for the variation that occurs in the project. There are several ways to do this:

- If your organization has a good data base of historical project actuals versus budget, you can estimate the average variation of actual versus budget cost and apply these data as factors to your estimate.
- You can total up the expected monetary value of your risk events.
- You can take an estimate of standard deviation of your critical path or estimate and multiply it by a constant (found in the Z-table in any standard statistics book) to determine a needed level of confidence.

- You can run your project through a Monte Carlo simulation using readily available software and use the results to determine your needed level of confidence.

Don't expand individual tasks to protect against risk; instead, have an explicit, traceable reserve.

Recall that a stakeholder risk tolerance assessment (part of step 6) is the idea that any internal or external stakeholder has certain comfort levels. Once the project manager has a good analysis of risk, he or she can negotiate with stakeholders in accordance with risk tolerance levels. Competent project managers explicitly consider stakeholder risk tolerances and include them in their baseline project plans.

For example, some stakeholders can tolerate reduced product performance but cannot tolerate delays. The plan identifies and quantifies the amount of buffer available for use by the project manager. Buffer is more accurately called *management reserve* and is the calculated expected value of negative risk events. This management reserve is a resource belonging to the project manager for absorbing negative risks. Baselining performance and the management reserve ensure ownership of risks and accountability for results.

Instruct the team to work through each risk and apply each response, even if the response does not seem feasible initially. You consistently will find that the team will adopt the proactive strategies of avoidance, mitigation, and transfer. If the team decides to accept the risk, it will have contingency plans for it. Residual risks are managed through a buffer established for that purpose or by accepting lower profits.

In your culture, there may be no alternative to padding estimates. Therefore, if you find yourself in a situation where there is no apparent alternative other than to pad the task, you must do it. If you are lucky enough to work in a culture that encourages good project risk management, you can follow the recommended practice of calculating an explicit and traceable risk reserve and then placing it into the project schedule or budget to buffer your project against risk exposure.

In step 4, we discussed classifying risks into categories. Often, the largest category is technical risks. We find that most technical risks involve product functions that do not perform perfectly. Thus, technical risks are really product reliability issues. They are risks only in the loosest sense of the word. Here is where a common product engineering technique—failure mode effects analysis (FMEA)—is helpful. FMEA involves identifying product functions and the performance of those functions, selecting failure modes for the function (e.g., intermittent failure), and assessing the probability of a given cause leading to the failure, its effect on the customer, and the ability to detect and correct the failure mode. FMEA is commonly taught in engineering schools and continuing-education programs, and there are numerous references to it on the World Wide Web, so many of the engineers on a team are already familiar with FMEA. The process we describe in this chapter is for program- and project-level risk; FMEA is better suited for technical reliability analysis.

Ideally, the team should assess each listed risk, but in practice, teams select a cutoff level and focus on risk events that have a higher score. Experience in projects indicates that teams generally will choose to consider 30 to 50 risk events but plan responses for only about 10 of them.

A good practice is to use the "top 10" list of risks and issues.

Effective risk management has its foundation in step 2, "Build Communications with Common Language." Recall that a risk is a discrete event with a probability and an impact, whereas issues are concerns that already have happened or are certain to occur. The team will get frustrated if it did not do a good job of distinguishing and understanding the difference between risks and issues. Because an issue is a certainty, the team is concerned not with *whether* the issue will happen but with *how* to resolve the issue to closure.

Management of project issues is straightforward. The team starts with the top-ranked issue and further clarifies it. Team members next determine an individual owner and establish closure criteria for the issue. We find it extremely helpful to assign ownership for *investigating* and making recommendations, often with a different person *deciding* on resolution of the issue. The project manager can keep team meetings productive by encouraging the offline resolution of issues and then using the team meeting to confirm the decision that the issue is now closed or needs to remain open. Experience with projects is that the team typically will chose to manage about 20 or so issues as a team and delegate the remainder to individuals.

Step 8: Integrate Risk Responses into Project Strategy and Document Project Baseline Commitments

Good project managers make commitments to their management to achieve certain results for scope, time, and cost. The project team has to make some decisions such as: What are the stakeholders' priorities? How much buffer should the team allocate to each project element (time, scope, and cost)? How does the team allocate ownership for the risks? Project managers are just like pilots who file their flight plan before starting, and they take into account risk events that might compromise their objectives for schedule, passenger comfort, budget, and safety. If the weather is severe, they may apply the response strategy of avoidance and cancel the flight because both the pilot and passengers have zero tolerance for accidents. In the language of risk, a weather-related accident is a low-probability, high-impact event.

Finally, we note that contracts (as we discussed in Chapter 4) are risk-allocation mechanisms. Therefore, all good project managers must have some knowledge of the different types of contracts used in projects so that they can understand the relationships between price, risk, and scope. The clarity of requirements is the starting point: More vagueness equals more risk. Firms set prices to cover the expected monetary value of the risks, as well as to cover costs and meet market demand.

Step 9: Execute and Control the Risk-Response Strategy

This step encompasses all the implementation work that the project team performs after the risk-response analysis and development, as described in steps 2 through 8.

Risk events and issues should have owners.

In the execution of the project, the project manager makes sure that the team is following its own processes. The team should have a prioritized list of risks and issues, a plan for how it will manage the risk events (if they occur) and how it will investigate and resolve issues, and what information it will record and archive. Team ownership of the project's risks is essential. Risk management is one of the very best ways to cause people to confront the difficult issues in the project and structure information so that people can make good team decisions. Communicating, team building, and decision making are processes, not events.

Discuss risks and issues in your project status and review meetings.

A good practice is to revisit the risk analysis and response planning periodically in the program. We suggest a weekly review of the risk and issues list, including asking the question, "What new risks or issues does the team need to be aware of?" Another commonsense practice is to formally reopen the risk analysis when team-defined triggers appear. Ask this question, "What triggers would stimulate a complete relook at the project?"

Recall that in step 4 we defined *strategic* risks and *operational* risks. Strategic risks are those that can lead to project cancellation, whereas operational risks affect the accomplishment of targets. In our opinion, a characteristic of world-class product development is a balanced approach to strategic and operational risk management. A good *gating* meeting focuses on *strategic* risk and recognizing need for early project cancellation. A project review or *technical review* meeting focuses on *operational* risks. It is a mistake to focus *exclusively* on operations: Keep a balance between strategic and operational risk.

While "flying your project," you should note these two learnings from pilots. The first lesson is loss of "situational awareness" and is often a team-based error. The flight team must know the location, altitude, and heading of the plane at all times. As an example, pilots on American Airlines Flight 965 chatted away as their airplane descended on the wrong flightpath and crashed into a mountain in Colombia on December 20, 1995, killing 159 people. The analogy is the *90 percent syndrome:* Projects appear to be on track until they are 90 percent complete, and then the last 10 percent of the scope takes the same amount of time as the first 90 percent. Complacency leads to loss of situational awareness and is a huge danger!

The second lesson is *target fixation:* a phenomenon in which military pilots become so narrowly focused on the target that they become oblivious to other things and crash their planes into an object because they are so narrowly focused on the target. We have seen similar experiences in projects when people become overly focused on the due-date target. Somewhat panicked, they declare, "We don't have time to plan!'' make unrealistic assumptions, make foolish compromises on quality, and burn themselves out with rework.

As project manager, you need to maintain your alertness and a sense of perspective. Project team members will take their cues from your actions. When risks and issues emerge, get the facts, and work on the responses. This is not a personal insult to you!

Step 10: Learning from Risk Management Activities

The final step in this team-based risk process is to review, reflect, and capture learning for future projects. Risk offers excellent opportunity for creating the organizational learning that undergirds product development. In the postimplementation review, the team members should consider these questions to stimulate reflection:

- Did we implement risk responses as planned? Why or why not? Were they effective? What can we learn?
- How effective was our issue-resolution process?
- Are we blaming individuals instead of looking deeper and more systematically into our organization and practices? What have we learned about our cultural preferences toward decision making?

BUILDING A CULTURE FOR GOOD DECISION MAKING

You now have a seen a simple 10-step method for building a team-based approach to project risk management. Risk is a more than a nice-to-do "tool." Because each project is different, its risks are different, and the responses must be flexible and robust. A sound risk management approach scales up or down to the innovativeness of the marketing, technology, and production strategy. In addition, the project manager should scale the amount of risk management process to the ability of the team to assimilate and use the process. Hence, the use of standard checklists and templates can focus the team's attention on the wrong areas and lead to a false sense of security.

Clearly, organizational culture influences how people will handle risk and issues management. Some project organizations stress telling the truth and working hard on problems. Others, regrettably, focus on blaming individuals even if the performance issues are outside the individual's control. This leads to conservative risk-avoiding behavior, and ultimately, these organizations may suffer.

Organizational culture sometimes conditions individuals to be distrustful. This distrust undermines good decisions. The fear that "I might be judged for something outside my control" often leads to avoiding participating in risk or deflecting the concerns to other team members. Conflict avoidance behaviors often reinforce the distrust culture. Conflict avoidance seems especially common in technical organizations. The pattern is this: People are busy and focus on their competency and their controllable area, which is a personal *comfort zone*. They avoid cross-functional issues because they fear appearing incompetent. Eventually, they can no longer avoid a given issue, and people react with anger and urgency, often blaming other people or organizations. A strange feature of this situation is that competency illogically equates to trust: If an individual perceives someone as lacking competency, he or she illogically concludes that he or she can't trust the other person. This gives the individual a reason to avoid further contact with that other person. Regardless, the urgent issue gets resolved, and people go back to their comfortable way of doing things. Because the root problems of avoidance and distrust don't get addressed, the cycle repeats. Risk management, because it is perceived as rational, detached, and objective, gives people a logical, pragmatic method for breaking the cycle. You can help people to improve their project management by encouraging active listening, by engaging people, and by coaching senior management on accepting the risk data.

Successful entrepreneurs and executives know that taking calculated risks is important to their success. No organization can make progress without taking risks. Risk management arguably is the single greatest competency of project management.

HIGHLIGHTS

- Risk involves discrete events that may or may not happen, whereas issues are certainties that must be resolved.
- Identify and evaluate your risk before you develop responses.
- Perform risk analysis and response planning as a team.
- There are four generic risk-response strategies: avoidance, acceptance, mitigation, and transfer.
- Analysis will identify both threats and opportunities.
- Most technical problems are issues; they will be resolved given sufficient time and money.

Part 3

Leading the People Who Work on a Project

12

Organizational Design for Delivering Projects

In this chapter, we describe three theoretical forms—functional, project, and matrix— by which organizations arrange their internal reporting relationships and chains of command. This understanding provides useful organizational context for making decisions on how to organize a specific project. We also describe issues associated with use of project management offices, venture teams, and other forms. Regardless of the formal organizational design to delivering projects, the informal organization plays an important role in successful project management.

As we discussed in Chapter 2, a project takes place in the context of a consuming organization and a delivering organization. Good project managers are aware of and understand the organizational structure of their organizations and of their customers' organizations, because the politics (the patterns of power and communication) will affect the performance of the project. This chapter will provide you with more knowledge of the organizational context, and that knowledge will help you to understand our recommendations for effective project teams in the following chapters.

THREE ORGANIZATIONAL FORMS

As we described in Chapter 1, projects are temporary work efforts that take the place of delivering organizations. The temporary nature of a project means that the project manager needs to understand the larger context of the both the consuming and the delivering organization.

In addition, projects frequently require the part-time use of resources, but permanent organizations try to use resources full time. Typical project requirements include the following: one hour of computer time each day for a week, use of a backhoe next Tuesday for the afternoon, one-quarter of Jane Draftperson's time this month and three-quarters of her time next month, and use of Joe Technician full time as soon as the project's circuit designer completes the design. No economically viable organization can afford to stockpile these resources to serve any project's needs instantly. Thus, it is important to organize for project work in adequately responsive ways, and it is important for project managers to recognize that this is a compromise that is not fully responsive to project needs.

Although no organizational form is perfect, it is important for the organization to support projects. This means that the organization must plan to accommodate this temporary disturbance and accept some disharmony.

> **Organizational forms differ in response to projects.**

Companies or their divisions or governmental organizations can be organized in a variety of ways and still manage projects effectively. The three most common of these organizational forms are functional, project, and matrix. In reading the following paragraphs, focus on the advantages and disadvantages of each of the theoretical forms.

Functional

Functional organizations (Figure 12-1) are common in companies dominated by business models that emphasize stable processes. For example, many manufacturing organizations turn out large lot sizes of a product. Most financial services organizations are oriented toward high-volume transactions that create a large amount of repetitive work. In these companies, performance metrics tend to emphasize efficiency of the process, and projects often are seen as exceptions to "normal" business. The person asked to manage a project in a company with a functional

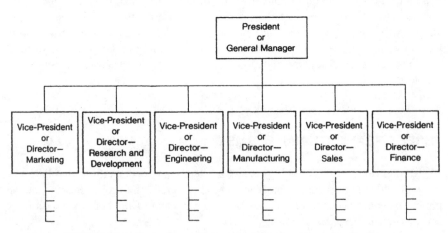

FIGURE 12-1. Typical organization chart of a functional organization.

organization generally has been oriented and loyal to the functional group to which he or she belongs. Specialists are grouped by function, encouraging the sharing of experience and knowledge within the discipline. This favors a continuity and professional expertise in each functional area.

Because such an organization is dedicated to perpetuating the existing functional groups, however, it can be difficult for a project to cross functional lines and obtain required resources. One project manager has observed that projects moved like a "lump in a snake" from one function to another in his company. Occasionally, there is hostility between different functions that shows up as barriers to horizontal information flow; open channels tend to be vertical, within each function. Such impediments to communication and cooperation produce what is known as a *silo mentality,* illustrated in Figure 12-2. The silo analogy is regrettably apt because destructive missiles tend to be launched from one silo aimed at another. "Siloed" organizations often have a hard time managing nonroutine projects. Absence of a project focal point also may trouble a customer interested in understanding the project's status, and functional emphasis and loyalties may impede completion. Because there is no focal point for the project, the president or general manager (who have far more to do than adjudicate cross-functional conflicts) takes on the integration responsibilities that normally would belong to the project manager.

From a project management point of view, the functional organization is least desirable.

The functional organization emphasizes functional skills by concentrating these in small groups. Thus, functional experts spend most of their time in proximity to and rubbing shoulders with people of similar skills. Unfortunately, this can isolate them from others with whom they must work if a project is to be successful. Many technical specialists, scientists, and engineers, for instance, often lack people skills.

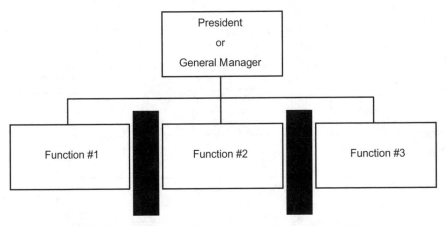

FIGURE 12-2. *Functional barriers can produce a silo mentality.*

This also may be true of other specialists, such as computer programmers. Because each individual function is best able to communicate with itself, interfunctional communication is often impeded and cooperation hindered.

However, if all the project's required personnel resources (including the project manager) are located within a single functional group, many of the problems can be avoided. Thus, when all project resources are located in one group, the functional organization may be a good choice.

Project

A project organization (Figure 12-3) is found in organizations that have unique customers and requirements. It is a common organizational form in the consulting, construction, and government-contracting industries. Line authority for a project is clearly designated, providing a single focal point for project management. All full-time personnel are formally assigned to the project, thus ensuring continuity and expertise.

A major difficulty with this kind of organization is the uncertainty these people feel about where they will go when the project is completed. This terminal anxiety can impede the project's completion. Career development of junior personnel may not be carefully or thoughtfully nurtured because a project's duration tends to be much shorter than an individual's career. There is also a tendency to retain assigned personnel too long. In addition, it is a rare project that actually has all the required resources assigned to it. Thus, such an organization still requires the project manager to negotiate with the remaining functional organization for much of the required support.

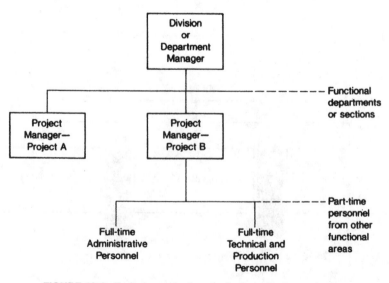

FIGURE 12-3. Typical organization chart of a project organization.

If the organization develops additional projects, managing them in this way leads to a splintering, with many separate project centers existing apart from the functional organization. Duplication of facilities and personnel can result. Managers within the functional organization may feel threatened as people are removed from their functional groups. This produces another series of stressors. Project organization often inhibits the development of professional expertise in functional specialties and may not effectively use part-time assistance from them.

> *The project organization form is most useful on large projects of long duration.*

The project manager must be a senior person (a so-called heavyweight project manager) to be effective in the project organization. Because he or she has direct responsibility for the work of everyone involved in the project, the project manager requires direct supervision of these people. Consequently, project managers in this organization are often of equal or greater rank than the functional managers. Good project organizations have an enlightened view about sharing power.

Matrix

The matrix organization (Figure 12-4) is a hybrid that may emerge in response to the pressures resulting from inadequacies with a functional or a project organiza-

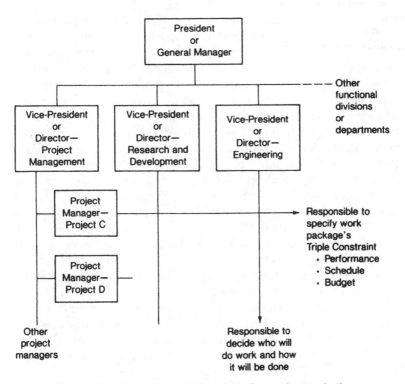

FIGURE 12-4. *Typical organization chart of a matrix organization.*

tion. It is a common approach to new product development in larger corporations. The matrix form is a response to the virtues of having functional groups but also recognizing the need to have a specific focal point and management function for each project. Line authority for a project is clearly designated, providing a single focal point. Specialists, including project managers, are grouped by function, encouraging the sharing of experience and knowledge within the discipline. This favors a continuity and professional expertise in each functional area. The matrix organization recognizes that both full-time and part-time assignments of personnel are required and simplifies allocation and shifting of project priorities in response to management needs. Functional departments are responsible for staffing, developing personnel, and ensuring the technical quality of the work done by those personnel. The project managers are responsible for defining the work to be done and establishing a reasonable plan (including schedule and budget) for accomplishing it. Project managers and department managers must agree jointly on tasks, goals, and the specific schedule for each of these.

> **The matrix is a response to complex demands for functional excellence and efficiency while responding to unique customer requirements.**

The main drawback is that a matrix organization requires an extra management function (namely, project management), so some managers perceive it as expensive and cumbersome. It is even possible to have a matrix organization within a matrix organization (e.g., the matrixed engineering department). In addition, the extra functional unit (i.e., project management) can rapidly increase bureaucratic tendencies, and the balance of power between project management and functional units can exacerbate conflicts.

A matrix organization may be either weak or strong depending on the project manager's power compared with that of the functional managers. The weak matrix may operate somewhat like a functional organization, and the strong matrix may operate somewhat like a project organization. This project manager's power may derive from financial control, seniority, or simply his or her persuasiveness. Thus, it is possible to have both a weak and a strong matrix within a single organization simultaneously.

Project managers in a matrix organization are often concerned about the work of tasks being done by people in functional organizations. Obviously, project managers wish to allow and encourage these people to have as much autonomy as possible to promote synergism, which should enhance the entire organization's efficiency and quality of work. This is particularly true where the functional groups have worked well in the past and have specific knowledge and expertise relevant to the task at hand. Conversely, the project manager is responsible for the ultimate result, must integrate each functional group's contribution, and thus wants to retain as much control and influence over each functional group as possible. Striking a satisfactory balance between these conflicting objectives is very difficult.

Operating in a matrix organization can be challenging when responsibility for a project or its key parts is divided or roles are not defined clearly. Project managers in this situation must get prompt resolution from management before the work direction is confused or the schedule is delayed.

Such managerial resolution may itself be contentious because of the inherent tension between the management of project and functional departments. Project managers often find themselves in conflict with account managers responsible for the external customer relationship. Account managers often want to develop customer goodwill by providing extra unpaid work, but this work will come out of a fixed project budget. It is clearly important to clarify expectations: Face and resolve this kind of problem early rather than hope it will happily solve itself.

OTHER ORGANIZATIONAL FORMS

The Project Management Office

Since the late 1990s, the concept of a project management office (PMO) has become increasingly popular as organizations realize the bottom-line contributions of successful project management. The PMO goes by a number of names, for example, the *program office* or a *project management center of excellence.*

There is no single "best practice" way for organizing and operating a PMO. At one end of the continuum is the "center of excellence" ideal, where the PMO acts as an internal consultancy providing project management support functions in the form of training, software, reference libraries, and other job aids. At the other end of the continuum, the PMO provides actual direct management responsibility for achieving the organization's objectives. In this case, top management has delegated authority to a specific PMO to act as an integral stakeholder and as a key decision maker. That is, the PMO becomes an organizational unit designed to centralize and coordinate the management of projects under its domain. In between these two extremes, the PMO may facilitate the selection, management, and redeployment of shared project personnel as well as dedicated personnel.

> *There are many ways to design a project management office.*

Quasi-Matrix

The quasi-matrix is also known as the weak or balanced matrix (Figure 12-5) and is a compromise way to obtain the benefits of the matrix form in a functional organization otherwise too small to afford it. In this compromise organization, when someone is designated project manager, he or she remains part of the functional group for project work done in that group (which is the same as in the functional organization). However, for project management work, he or she reports directly to top management.

Thus, the project focus has top management's support to cut across functional group boundaries. In some cases, the quasi-matrix will have a manager of projects (just as in the matrix organization), rather than the boss, to whom the project managers report for project (as distinct from functional) matters.

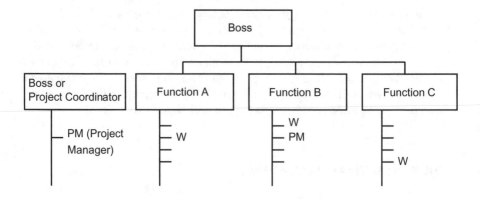

W = Worker or project contributor

FIGURE 12-5. Quasi-matrix organization.

Unlike the "heavyweight" project manager required in the project organization, a "lightweight" project manager may suffice in the quasi matrix. In many situations, the role can be filled with a middle- or junior-level person, often a technical specialist (such as a design engineer, programmer, and so on). Many of the project contributors continue to report to their functional managers, and the project management role may entail primarily coordination.

Venture

The venture organization, common in several very large, commercially oriented companies, is especially appropriate for projects aimed at new product development. Basically, the goal is to set up a tiny functional organization within a giant corporation, thus achieving the advantages of compact size, flexibility, and the entrepreneurial spirit of the small company within and supported by the financial, physical, and human resources of the larger company. Where such a management organization exists, it is common to team up an engineer with a marketing person and a manufacturing person in the earliest phases of new product development. As the effort moves forward, the venture organization grows, ultimately becoming a functionally organized division within the parent company.

Task Force

A task force may be helpful in coping with an unexpected project.

Organizations frequently use a task force to cope with an unexpected project. Hence, this response is used most commonly by a functional organization because the other organizational forms are already able to deal with projects. A task force may be thought of as a rarely used single-project organization within the functional organization. It can be formed quickly, usually by a very senior officer.

Although the people selected to serve on a task force may be highly motivated by their selection, they frequently are not relieved of their usual duties and thus may not have sufficient time for the task force. If they are relieved of their normal duties, they may be anxious about their assignment when the task force has completed its job.

Other Project Teams

Today, other labels are applied to various special-purpose project teams. These include *core teams, concurrent engineering teams, reengineering teams, benchmarking teams, process development teams,* and *SWAT teams.* We are sure other labels will be in vogue in the future. In most cases, such a label merely describes the role of an *ad hoc* project team.

Some teams may be self-managed, although this is often more in the label than in reality. Assignment to such an alternative project team may be full or part time; the latter is more common for all but emergency situations.

THE INFORMAL ORGANIZATION

Regardless of the formal organizational structure, there is always an informal organization (Figure 12-6). Where friendships and common interests are present, these create channels through which information flows easily, and cooperation is

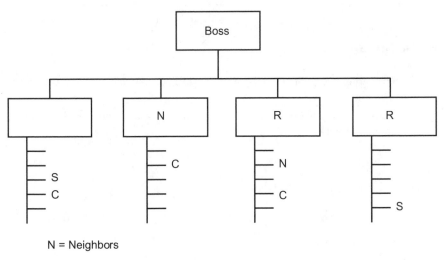

N = Neighbors

R = Rivals

C = Worked cooperatively together on previous project

S = Spouses are friends from college

FIGURE 12-6. *Informal organization.*

encouraged. Conversely, rivalries and animosities can inhibit cooperation. In such situations, the enemy of my enemy may still not be my friend. Newly or recently hired employees are at a disadvantage because they are unfamiliar with this informal organization, and choosing such a person as project manager can lead to difficulties.

TYPICAL PROBLEMS

Each form of organizational structure has its advantages and disadvantages, and it is common to encounter hybrids. The only real problem occurs when a project manager believes that a different organizational form will solve all the organizational problems he or she is experiencing. In fact, no organizational form is perfect for all situations or for all time. Thus, reorganization is required regularly in any organization doing project work to keep the organization as useful and productive as possible.

The persistent problem is carrying out any project that crosses the lines of more than a single functional department. In these cases, the project is likely to be at the mercy of the functions. As one project manager stated, "Each functional department determines its own priorities and staffing needs." When a project is suffering because of this parochial attitude, the manager must first attempt to resolve the problem directly with the department manager. Failing that, he or she must bring the problem—and some proposed solutions—to the attention of senior management.

HIGHLIGHTS

- Three common organizational forms for project management are functional, project, and matrix.
- The informal organization is always present and affects how well the formal organization works.

13

Building the Project Team

Team building is one of the most important functions of the project manager. In this chapter, we discuss sources of project personnel and consider the necessity to compromise by using whomever is available. Then we deal with how much control a project manager can exercise over project personnel and provide tools to help him or her manage the project. The next section discusses the use of task assignments as a means both to assign the work packages and to obtain commitments from personnel to carry out the work. We also discuss the virtual project team and using scheduling software in a team.

There is a sarcastic project management quip that goes, "The next time I have a project team, I'm not going to have any people on it!" Dealing with people is one of the most frustrating elements of a project, yet one of the most incredibly rewarding experiences in projects is developing relationships—and even friendships—as the team accomplishes remarkable project results.

Of the many important activities of successful project management, the greatest are the processes of recruiting, building, and maintaining a project team. Almost by definition, individuals have variation, and their personal differences lead to different bundles of strengths and weaknesses. If you get the right blend of people for your project, the project will go amazingly well. If not, you may be in for an unpleasantly stressful endeavor. However, just recruiting the best people does not ensure that your project will be successful. You may have heard the saying that all-star sports teams are tough to manage and world champion teams are seldom composed of all-stars. A good manager knows how to deal with each individual's moods and skills and get the best out of everyone.

Get the right people, and get the people right.

CORE TEAM AND EXTENDED TEAM

If you are working on a complex project that has an aggressive timeline, you should consider the core team and extended team model of project teaming. The *core team* is small group of people (typically fewer than 10 individuals) who provide a cross-functional, cross-discipline view of the project. The core team's membership is stable across all phases of the project life cycle, and this stability decreases the probability that there will be miscommunication during a handoff and provides more of a sense of individual commitment to the results.

The *extended team* is composed of the people who perform much of the "heads down" work on the details of the work. They channel their work to a person who is a member of the core team. Here is an example of how an extended team member and core team member might interact:

> Helen is marketing specialist and is a core team member for a new product launch. As a core team member, Helen must contribute to all elements of the new product development strategy for that product, and she realizes that the team will be looking to her for leadership in developing the new product pricing, promotion, and distribution strategy. One specific task is to develop an illustrated product brochure, a task that will require a special skill. Cindy is a graphic artist and an extended team member who has the skill to develop the brochure and works in the same department as Helen. Helen will bring the assignment to Cindy and (with the knowledge and approval of Cindy's boss) will explain the requirements for illustrations to her. When Cindy finishes her illustrations, she will deliver her work to Helen, who will take it back to the core team for integration with other work efforts.

In this example, Helen has assumed some of the project management responsibilities. Other core team members are performing similar interactions with other extended team members.

Notice in the preceding example that there is no mention of the project manager. Because of the fast timing and complexity of the project, the project manager is focusing on integration and strategic issues, for example, the risk and issues management that affect the team's ability to deliver a commercializable product to the customers on the target launch date for the target price. We will expand on the competencies, roles, and responsibilities of project managers in Chapter 15.

STAFFING STARTS WITH PROJECT SCOPE

Human resources management is an important project management knowledge area. When staffing a project, use the following order of analysis and action:

1. *Requirements.* As we have discussed, customers and their requirements are the fundamental reason for a project; thus, the discussion of staffing *must* start with the work to be done (the work package) and the expectation for when it will be done (the critical path).
2. *Competencies.* Specific work activities require specific skills. Since certain people can do certain things, the fulfillment of requirements is linked to the skills possessed by people.

3. *Availability.* Once you know what skills you need, you have to determine how many individuals you need, their level of involvement (full time or part time), their location, and so forth.

When it comes to staffing projects, project managers and functional managers typically have different priorities. Project managers' first priority is delivering the work that satisfies requirements. On the other hand, functional managers typically are measured on how efficiently they use their annual budget. The conflict between project managers and functional managers often can be explained as follows:

- The project critical path determines the sequence of activities. The very nature of the work activities creates dynamic requirements for staffing; on any given day, the number of staff may be greater or less than the capacity needed for the preceding day. While resource leveling is an option, the usual result is to extend the duration of the project.
- Projects are measured by *effectiveness,* meaning the delivery of results to the customer.
- Functional managers are measured by *efficiency,* meaning keeping people busy and getting the most they can out of their annual budget. For them, it is desirable to assign underutilized people (even if they are not the right match for the project requirements).

FORMAL PROJECT AUTHORITY

In Chapter 2, we wrote, "If you get the authorization step right, you are much more likely to be successful in your project." Good organizations formally authorize projects through charters or other types of contracts. These charters spell out the formal authority of the project manager. For example, the charter would explain that certain people "re-port"—for purpose of the project—to the individual named as project manager and not to their traditional supervisor. The charter also should describe the authority of team members to speak for their department, which overcomes the common problem of functional managers overriding the decisions and commitments that people from their department may make to the project.

> *The project charter describes formal authority of individuals.*

ASSIGNING PERSONNEL TO THE PROJECT

Table 13-1 shows eight categories of personnel assignment to projects. They result from all possible combinations of three factors: (1) whether personnel report directly to the project manager or are administratively assigned to someone else, (2) whether they work full time or only part time each day (or week or year) on the

TABLE 13-1   The Project Team and the Support Team

Duration of Project Assignment	Nature of Reporting Relationship	Reports to Project Manager		Works on Projects but Reports to Another Manager	
		Works Only on Project	Also Has Other Assignment(s)	Works Only on Project	Also Has Other Assignment(s)
From start to finish		P	P	S	S
Only a portion of project's duration		P	P	S	S

P = project team.
S = support team.

project, and (3) whether they work on the project from its inception to completion or for only some portion of the project.

Project Team

The project team is composed of the people who report administratively (directly or through other direct subordinates of the project manager) to the project manager (the four cells designated with a P in Table 13-1). The project manager can assign work packages to these people rather than having to negotiate with other managers to obtain commitments for their work. (The support team is composed of all other personnel working on the project, designated with an S, including subcontractors.) The core team may be composed of both project and support team personnel.

Team Matrix

The amount of project labor obtained depends on the delivering organization's project organizational form (i.e., functional, project, or matrix) and project size. In a matrix organization, contributors will come from all areas of the organization. In a pure project organization, the vast majority of project labor, perhaps all of it, may be assigned to the project manager. This is especially likely for a large project of long duration. Small projects are not likely to have their own personnel, regardless of organizational form.

There are eight ways to assign personnel to a project, so the project manager must manage differently.

A key point that emerges from consideration of Table 13-1 is that the project manager must provide eight different kinds of management attention to people working on the project. People who work on the project for only a portion of its duration must be managed to be ready when needed; then orientation to the project must be provided. Finally, the project manager must recognize that they may be frustrated or lack a sense of accomplishment at leaving the project prior to completion. People who have other assignments must be persuaded that the work they do on the project deserves their attention each day (or week). If everyone who is working on the project is doing so on a part-time basis, other work (sometimes more interesting but of lower priority) intrudes. Because they may have a lower stake in the project, they often require better or more forceful leadership. People on the support team are influenced by other managers, which can lead conflicting priorities and direction. Issues of priorities, performance standards, and loyalty often require the project manager's attention.

SOURCES OF PERSONNEL

There are many sources of people, including the proposal team, other people already employed by the organization, and people from outside the organization (e.g., hired personnel, contract personnel, consultants, and subcontractors).

The Proposal Team

A very good source of project personnel is those who worked on the proposal. They are familiar with the customer's requirements and the technical approach to satisfying them—they are already "up to speed." They will, for example, understand the meaning of potentially ambiguous words (such as *security*). In many instances, they already will have "bought into" the project. Although the proposal team may be highly qualified, it may have insufficient capacity for the project.

The proposal team is one good source of project personnel.

Other Organizational Employees

Other employees of the organization are another source of personnel. These people are at least familiar with company policies and procedures; they know where the library, the model shop, and such are located. Although they may not be familiar with your project's requirements or the proposed technical approach, they are at least familiar with how the company does business and know its strengths and weaknesses. They know whom to call for help and where to go to get something done. In fact, they probably have worked on similar projects in the past.

The project manager may know their strengths and weaknesses and thus be able to assign them to appropriate work packages.

People from Outside the Company

There are a number of outside sources for personnel (Table 13-2). Consultants, contract ("body shop") personnel, and subcontractors can be obtained quickly.

To hire a person, a project manager has to have a personnel requisition approved and typically must advertise the position, interview several people, make one or

TABLE 13-2 Filling Personnel Needs

Time Duration		
Permanently	Permanently or Temporarily	Temporarily
Hire	Transfer	Consultant
		Job shop
		Subcontract

more offers to get an acceptance, and wait for the person to relocate (if required) before coming to work. Then there is an orientation period while the person becomes familiar with company practices. Because it often takes months to hire a new person, a project of short duration rarely can afford the time to hire personnel and thus depends on the support team.

COMPROMISE

It is rare that a project manager can staff a project entirely with personnel who (1) already work for him or her, (2) worked on the proposal, and (3) comprise exactly the right distribution of skills to carry out the project. Usually, the project manager must staff the project from whomever is currently available either full or part time. Many of these people will not completely meet the staffing requirements. It is often a case of fitting square pegs in round holes. Thus, when you plan your project, plan the schedule and cost using realistic assumptions.

Staffing compromises are usually necessary.

Qualifications

Sometimes functional managers in the company assign their underutilized people (which often means that they are lesser qualified) to the project. These people may be marginally employed, so company management may pressure the project manager to accept them into his or her group. Consider the problem one project manager cited:

> I inherited an employee who had been with the company for approximately 35 years. In all but one or two years of that time he was never properly supervised. He did, however, receive regular raises. He's apparently been allowed to set his own standards of performance, but these are subpar. He has proven to be very devious at beating "the system" with regard to hours and personal activities that are not job related. Now, my attempts at corrective actions and discussions with him produce a very defensive position. He says, "I never knew I wasn't up to par." He exhibits resentful behavior, including mistakes and sulking.

In this situation, there is pressure from above to accept the people, and there is another manager offering them as freely available. However, if these people are unproductive or unqualified workers, it is probably better to terminate their employment than to shift them from one project to another. Nevertheless, it is common for a newly appointed project manager to be offered all kinds of personnel for transfer. On a short-duration project, it may be better to accept these workers, unless they are clearly unqualified, than to recruit better-qualified assistance.

Motivation

Some projects that offer high pay (such as major construction projects requiring a great deal of overtime or shift premium work) frequently attract workers whose primary motivation is money. The project manager may be besieged by candidates who wish to go to work on his or her project. Their motivation, however, may not be best for the project. The manager should seek to staff the project team with a few high-quality people and confine the money seekers to support team roles, where they are someone else's problem.

Conversely, a project with high scientific content or one of national importance (such as a new medical instrument system that reliably diagnoses any cancer at a very early stage to permit effective treatment or, perhaps in the future, a manned landing on Mars) often attracts highly dedicated, altruistic people. A common correlate of this altruism is a lack of practicality, which the project manager must watch for and temper.

The project manager needs to develop skills in motivating and coaching team members.

Some projects have an unsavory reputation (fairly or unfairly earned) that makes it very difficult to recruit personnel. They often require portions of the work be performed at an unattractive or remote location. To overcome this drawback, various inducements may be required.

Recruiting Qualified Help

Some compromise is required in staffing the project team, but there may be some skill requirements that cannot be compromised. Most project managers prefer to have people on the project team because it seems to improve project control. People on the project team cannot be given other distracting work assignments unless the project manager approves it, but people on the support team may be given other work that detracts from their ability to honor support commitments. Thus, where you want the best-qualified help for your project, you must locate a person with the required skill and try to obtain a commitment.

Although it is not always possible, try to get people to volunteer for your project. Experience shows they will be much more motivated than those who are assigned (perhaps arbitrarily). Many project managers have found value in volunteering for project management positions with their church or a civic organization because they have had to learn how to recruit and motivate people in a setting where there is no formal, legitimate authority for the project manager position. People will volunteer for a project when they find the project supports their personal interests and values.

CONTROL

Supervision

The people who staff a project probably will change. They may not work full time on the project, either for its entire duration or full time within any given workday or workweek. Nevertheless, they are under the project manager's direct supervision unless there are intermediate levels of supervision. Some of these people may have been transferred from other managers.

Projects go through different phases, which implies that personnel may change. For instance, the creative design person, so valuable in the early phases of system design, may not be needed when the project is moving toward completion and the team is trying to finish what has been designed rather than figure out additional clever ways to design it.

People will join and leave the project as needed.

Thus, some project team people may have to be reassigned during the project. An administrator or junior project manager assigned to work under the project manager may be needed the entire time, but other personnel may need new assignments. They will either go to work on another project full time after completing their work or work on two or three projects part time. It is therefore important that the project manager exercise control over the timing of these assignments to have people with the right skills available when required and have other assignments for them when they are not required. This is one reason a resource-allocation analysis is desirable.

Proximity

One of the project manager's most powerful tools for improving project performance is *colocation,* that is, locating project team members close to each other, typically in a common area. This aids communication, and where there is increased communication, there is increased understanding.

Colocation is most appropriate for projects of long duration requiring a large team.

Colocation is often impractical because the people from a key required function are still needed in their "normal" (nonproject) job or some key functions are located in multiple other facilities (e.g., dispersed manufacturing sites). Figure 13-1 summarizes some of the project duration and team size issues in deciding whether colocation is appropriate for your project.

Team Size				
	Large	Intrateam communication "overhead" costs are higher. An investment to promote teamwork is required. A more experienced project manager is required.	It frequently reacts to marketplace changes. Miscommunication, confusion, inefficiency, and wasted money are possible.	Team co-location is often required and justified.
	Small	Generalists may be required. A less experienced project manager may be adequate.	Cost normally is low.	It must be determined if a larger team could shorten the schedule.
			Contributions of part-time people frequently are required. Team colocation is hard to justify. Changing the project manager is normally not practical.	The project manager probably will change as phases shift. Full-time, dedicated project team members can be used. Boredom or burnout are risks. Team colocation is possible.
			Short	**Long**
			Project Duration	

FIGURE 13-1. *Team organizational issues affected by team size and project duration.*

Stability of Team Membership

Especially if you are using the core team–extended team model described early in this chapter, you should try to maintain consistent team membership and not allow people to come and go in the project. One large company did an analysis of its projects and found that every time there was

> *Try to keep the team membership stable.*

a change in the project manager, the project suffered a three-month delay. Every time there was a change in a core team member, the project suffered a one-month delay. In knowledge-intensive projects such as software or new product development, there is a phenomenon known as Brook's law, which is, "Adding people to a late project will make it later." The reason for the loss of productivity on knowledge-intensive projects is that people who are knowledgeable have to take time to bring the new people up to speed.

The disadvantage of this recommendation is that the project manager will have to contend with many personnel problems, such as people quitting (either the company or the organizational unit) to work elsewhere, sickness, reassignments made by higher management, lack of interest in the project, or other conflicting assignments.

TASK ASSIGNMENTS

We have previously described the use of the work breakdown structure (WBS) and network diagram to divide projects into small pieces of work. Each of these pieces, or tasks, has a corresponding cost estimate. In the ideal world, the person responsible for each task has participated in preparing both the schedule and the budget estimate. This person also should have played a significant role in planning his or her work package. In any case, the project manager must assign tasks to many different people.

> *All work assignments should be written.*

The project team member who now has his or her task assignment should provide the project manager with a detailed plan of how that task will be performed and periodically review progress against the plan. To the extent that the task performer has played a major role in creating and initiating the task assignment, he or she is likely to be highly motivated to carry it out. Conversely, if the task was assigned without negotiation, the person may have a low sense of involvement and may be largely demotivated by the assignment.

There are five substantial benefits in structuring the WBS so that task assignments are small:

1. A small task usually has less risk than a large task.
2. The goal of a small task is more likely to be clearer than the goal of a large task.
3. There will be a greater sense of urgency because the end date is closer. (In a famous experiment, Amos Tversky and Eldar Shafir offered students $5 to complete a long questionnaire. In all 66 percent of those with a 5-day deadline collected, 40 percent of those with a 21-day deadline collected, but only 25 percent of those with no deadline collected.)
4. There is faster psychological reinforcement and increased motivational impact (assuming that the task is completed).
5. You get early warning of difficulty (if the task is late or not completed at all).

THE VIRTUAL PROJECT TEAM

Increasingly, organizations are using the virtual project team model. The distinguishing characteristic of a virtual project team is that people spend little or no time meeting face to face. In many cases, virtual project teams have members working on different continents!

Organizations have found a number of advantages with virtual team project teams, such as the following:

- They can better leverage the knowledge and skills of people from the same company who live in widespread geographic areas. This includes stakeholders such as suppliers and users.

- They can include people who work from a home office or who work different shifts or hours.
- They can better accommodate people who have disabilities such as blindness, deafness, and lack of mobility.
- They can invest in projects that the organization otherwise would not perform owing to travel expenses.

The use of virtual teams has grown with the availability of enabling technologies such as e-mail, Web-enabled project management software, cell phones, instant messaging, overnight package delivery, fax, personal digital assistants, video conferencing, and the like. These technologies make it possible for people to speed the transmission of requests and information; for example, the use of instant messaging allows a person to ask another person a question and get a response within seconds.

The technologies also improve the ability of the team to work asynchronously; for example, a project manager in North America can authorize (by e-mail) a work package to a software development team in Asia at the end of her work day, and the development team can start that work package and have it back to the project manager at the end of their work day (which means that the completed program is in the project manager's in-box when she arrives for work the next day. All this work has transpired in the course of a 16-hour period!

If you are in a virtual project team format, don't fall into the common trap of *instrumentalism,* that is, treating people as only a means to get a task accomplished. The transformation of the workplace in the past decade has been remarkable, but people's cognitive ability and emotional needs have not changed. The best project managers of virtual project teams maintain (and even increase the frequency of) their use of the "soft skills" of rapport building, empathy, and authentic respect for their teammates. The goal in having a virtual team is to have the same kind of "feel" as a face-to-face collocated team. This goal puts more demands on the project manager to facilitate and lead communications.

Here are a few tips in for building and managing a virtual project team:

- Meet face to face at least once, preferably early in the project and most effectively at the kickoff meeting (regardless of travel cost and time).
- Develop ground rules about use of the different technologies.
- Spend relatively more time on agenda building and dissemination before meetings.
- When using teleconferencing or videoconferencing, have each person announce himself or herself before speaking.
- When working across multiple time zones, "spread the pain." Don't repeatedly ask the same group to be available at 2 A.M. local time to participate in the meeting.
- Remember that body language also communicates a substantial amount of information, and in certain "high context" cultures, the body language is even

more important than in others. We have found it useful to say, "I can't see your face. Tell me what you're feeling," if there is a potential for misunderstandings.

- Summarize and document your discussions. Take advantage of the various Web-posting boards to keep "everyone on the same page."

TURNING A GROUP INTO A TRUE TEAM

People casually use the word *team* in projects to refer to the people who are working on the project. By using the label *team,* people hope to get cooperative, teamwork-type behaviors. However, based on our experience, it is rare to see a project being performed by a true team; instead, we see groups of people complying with directives, doing what they can to get by, and cynically regarding their peers.

There are many books and courses on leadership and team building, and if you consider what they have to say, you will agree that a team is a rather unique, special kind of environment that is characterized by trust, mutual accountability, common purpose and methods, and high performance. (If you've ever been part of an outstanding sports or work team, you will know what we mean. There is something intangible and special about the experience.)

Building a true team takes an investment of time and emotional energy and is a skill. Many people talk of the following process:

- *Forming.* Recruiting people so that they agree that they want to be a member and in some way personally identify with the group of which they are members.
- *Storming.* Recognizing that different people have different values and styles of doing work and that conflict needs to be recognized and managed.
- *Norming.* Developing a consensus around the rules that the team will use to surface issues, make decisions, and accomplish its work. When groups fail to become teams, it is commonly because they cannot effectively develop team norms.
- *Performing.* Achieving a high level of performance.
- *Adjourning.* Because the project comes to an end, the team comes to an end. If you have been part of a true team, you will recall that there is a certain amount of sadness as you realize that you are not likely to work with the same group again. Most true teams have some sort of closing meeting or party to acknowledge that people are moving on to new assignments.

COMPUTER SOFTWARE

Figure 13-2 illustrates how one software package (Microsoft Project) can facilitate team (both project and supporting) organization. Columns are created for each

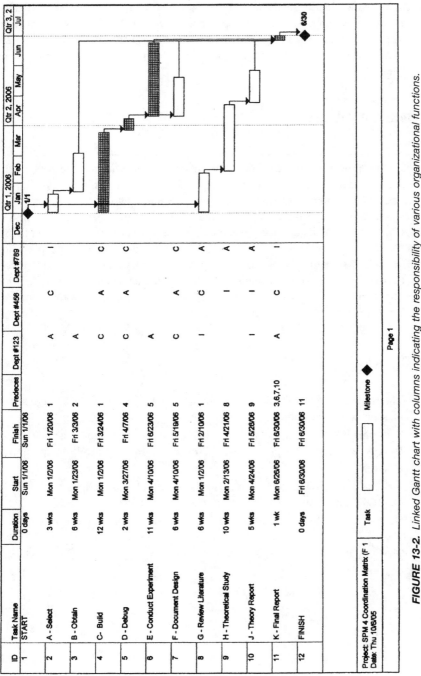

FIGURE 13-2. Linked Gantt chart with columns indicating the responsibility of various organizational functions.

function (or individual) involved, and the nature of their involvement is indicated. The figure shows three illustrative departments (Dept #123, Dept #456, and Dept #789). Each task must have one function (or person) that is accountable (A) and may have other functions for which concurrence or input is required (C or I, respectively).

TYPICAL PROBLEMS

The usual problem is what to do with unqualified or unproductive personnel. This is one reason you should follow the risk-response planning discussion in Chapter 11. Because sometimes you have personnel who must be used in an area outside their specialization, you need to provide budget and time for training. In the case of unqualified personnel, you simply can refuse to accept them on the project team. You also must expect to work with a group of people with differing and sometimes conflicting personalities. Obviously, these people must work together effectively. Depending on the specific situation, you may have to solve individual problems, confront or replace people, or work with the entire group to improve morale and cooperation.

HIGHLIGHTS

- The project core team is a small group of people who are focused on decision making and have stability throughout the project.
- It is important to understand how your organization authorizes projects and assigns authority. The project team "reports" administratively to the project manager.
- Sources of project personnel include the proposal team, others employed by the organization, and people from outside the organization.
- Having team members in close proximity improves communication and hence control.
- The virtual project team is increasingly common and offers special leadership challenges.

14

Organizing the Support Team

The support team, sometimes called the extended team, is composed of the people who work on the project either full time or part time for a part or the entire project. This chapter discusses how to obtain their involvement and commitment and how their efforts can and must be coordinated with the project team. We also consider interaction between the project team and support groups and subcontractors.

INVOLVEMENT AND COMMITMENT

As with the project core team, a good way to develop a sense of involvement and obtain a commitment from the support (or extended) team is to have had its members participate in the proposal. Participation also builds a team spirit that continues beyond the project. Failing this, their involvement in planning their own work and committing those plans to writing also should elicit involvement and

Involve support team members in the proposal phase or as early as possible.

commitment—remember the golden rule introduced in Chapter 5: Get the persons who will do the work to plan the work.

Project managers, regardless of the specific organizational form of their company, often feel they lack authority. In a project organization, project managers do have hierarchical authority over the people working for them. Nevertheless, in all situations, as you will see in the next chapter, formal authority per se is not especially useful. Influence is an important type of organizational power.

Early Support Group Involvement

Project managers and the project team often ignore support requirements, which other groups must provide, until it is too late. This is one reason why it is common to call them the *extended team:* Individuals can recognize that they have membership in the project and a stake in its success. In larger corporations, support groups include such functions as purchasing or procurement, stores and inventory, shipping and receiving, contract administration, documentation, model shops, other specialist services, and so on. Unless support personnel understand that their services may be required, they cannot anticipate the extent to which they will be needed. When support is sought tardily, support groups feel left out, and it may be difficult to obtain their commitment.

This kind of situation may arise because the project team has some degree of parochialism or is not aware what support is readily available. The project team may not understand the potential roles others can play or may assume that it knows better than the support groups what kind of effort will be required. This latter situation frequently arises because the project team feels that a support group will "gold plate" the amount of work they propose to do, exceeding project budgets.

As stated earlier, these problems can best be mitigated by involving support groups in the proposal phase. If this cannot be done, involve them as early as possible in the project work. Give them an opportunity to participate in planning their task and employing their best thinking and expertise.

Have support team members estimate the time and cost of their tasks.

The same applies to the time and cost estimates. The support group should make time and cost estimates for their task, and the project group should approve them. These estimates may require a negotiated revision to adjust other project tasks to accommodate support group plans if they differ from the project team's first estimate. This is a common occurrence. Support groups sometimes must perform their role at a pace dictated by other, higher-priority commitments, thus scheduling your project support differently than you had planned. Sometimes the support group sees a completely different way to undertake its role, often to the project's advantage. Or the support group's experts may convince the project team that their role must be broader than originally conceived. For all these reasons, involve support groups as early as possible.

Written Commitments

Document your agreements for the effort and deliverables from supporting groups.

Obtain meaningful commitments from support groups within your organization just as you do from outside subcontractors, namely, a written agreement. (This is also what you should do with project team member commitments; the only difference is what actions you can take to settle disputes that may arise.) If the support group manager must sign a written agreement, he or she will be motivated to make his or her group live up to its commitment. As we described in Chapter 2, verbal contracts are legitimate and enforceable, but it would be better to listen to the voice of experience and document important understandings. An observation attributed to Samuel Goldwyn is highly pertinent: "A verbal contract isn't worth the paper it's written on."

Support Team Advantages

As we said in the preceding chapter, most project managers seem to prefer to staff their projects entirely (or mostly) with project team members. However, a project manager (especially in a matrix organization) might prefer to have a large support team rather than a large project team for the following reasons:

1. The project manager does not have to worry about the support team after the project ends.
2. In the case of subcontractors, the support agreement is embodied in a legally binding instrument, namely, the subcontract or purchase agreement.
3. The project manager has the whole world in which to find specialists or experts with the required skills.

COORDINATION

Once the support groups have been identified and their work has been planned properly and phased in with that of the project team, there is a continuing need to coordinate project teamwork. This is best done with network diagrams (Figures 14-1 and 14-2). In both figures, support group work has been segregated from the main part of the network. There are many other ways to do this, for instance, using distinctive line patterns for each support category. Where color copying machines or printers are available, a color code may be used advantageously. Some computer-based project management software allows you to generate schedules for specific groups or individuals. Other software packages allow you to dedicate portions of the schedule chart (e.g., the upper two inches, the middle three inches, or the lower two inches) to the work of a single group or individual. (Figure 13-2 shows an alternate way to facilitate coordination.)

A network diagram can aid coordination.

Change

Consider the network diagram of Figure 14-1. Imagine that you discover that you will be late on task D. Should you inform your subcontractor (on task B) of this lateness? It depends. In general, it is probably best to advise subcontractors of your true need date. If you do this, you make it easier for them, and their costs to you will be lower (at least in the long run). However, if subcontractors have a history of lateness, it is probably best (1) to have originally allowed time contingency for their work and (2) not to let them know of any delay you have experienced.

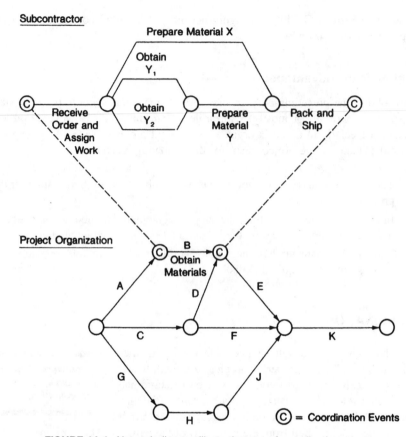

FIGURE 14-1. Network diagram illustrating use of coordination events.

Communication and coordination primarily should be in writing. Change, when required, should be accomplished by prompt oral communication, over the telephone and/or at meetings involving as many people as required. Then you should ensure that you record the change in the project baseline.

Revision

Write and distribute plan revisions.

Once committed to paper, plans must be disseminated to and understood by all involved personnel. Plans also must be maintained in a current status. If any out-of-date project baselines are allowed to remain in circulation, the credibility of all project plans will become suspect. Therefore, everyone who had the original plan must receive revisions. This can be facilitated by keeping an accurate distribution list. Where your company has a communications network, each person

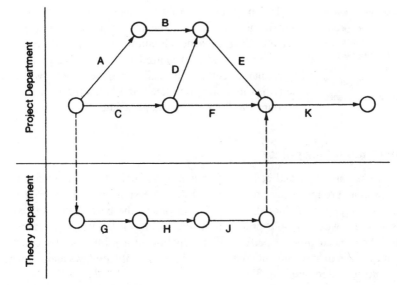

FIGURE 14-2. Network diagram illustrating use of spatial segregation of a support group's activities.

(or at least each key person) may have access to the current schedule. Obviously, they must both consult and understand this schedule and any changes.

INTERACTION WITH SUPPORT GROUPS

Project team and support group interaction can be difficult. All too often, the support group is brought in too late, a situation that recalls a story. A commuter comes dashing onto the train platform just as the morning train into the city pulls out. A bystander, observing that the commuter has just missed the train, comments, "Gee, that's too bad. If you'd simply run a little bit faster, you'd have caught the train." The commuter knows, of course, that it is not a matter of having run a little bit faster but rather of having started a little bit sooner.

Project Actions

The tardy commuter's situation classically applies where the purchasing department is involved. Purchased materials arrive later than required, the project is delayed, and project people blame either the subcontractor for delivering late or the purchasing department for failing to place the order early enough. In fact, the blame lies with the people who did not requisition the purchase sufficiently early so that the goods would be delivered on time. They need not run faster; they should have started earlier.

Start support group tasks early enough so that they can be finished when required.	The experienced project manager copes with this problem in two ways. First, he or she makes certain that the network diagram schedule allows enough time for the support groups to perform optimally. Second, the project manager makes certain that all personnel know when task activities must be completed and holds the task managers accountable for meeting the schedule.

Support Team Viewpoint

These issues can be looked at in a different way, namely, from the point of view of the support groups. They are composed of professionals, in the preceding example, purchasing professionals who wish to obtain the best quality of required goods at the lowest possible price within the other constraints project personnel impose. They need time to perform their function in a professionally competent way. In the case of government contracts, there commonly are laws and regulations that require three competitive bids.

No job is too hard for the person who does not have to do it.	The manager of a project support group one of us worked with observed that some project teams failed to take advantage of the support group's expertise and knowledge of its existing workload commitments. The support group manager quoted in Chapter 5 illustrates a frequent complaint that their expertise and knowledge is not considered

early enough to be helpful. For instance the already existing workload imposed on the support group at the time when a project needs their participation may preclude timely assistance.

Support groups labor under many constraints.	The same is true, of course, of any support group— technical writers, computer programmers, designers, draftpersons, or model shop personnel. Everyone wants to do a good job and wants sufficient time in which to do it. However, departments have a workload imposed on them by

others. They are trying to respond to many projects bringing work to them at random times and in variable amounts. Thus, support groups typically have some backlog they must work through before they can get to new requests. If they did not have this backlog—if they were sitting there idly waiting for the work to arrive—they would not be using a vital organization resource, their own time, in the most effective way.

SUBCONTRACTORS

Subcontractors are basically no different than you. They have a contract from a customer, in this case, you or your project. They want to be responsive to you, but they have the same kinds of problems you do: Personnel and resources frequently are not instantly available or perfectly suitable; they need time to plan their work; they have to balance competing demands; and so on.

Just as a contract controls your relations with your customer, so subcontractors define their relationship to your project by the contract your company's purchasing department issues to them. Should a change be required, it is certainly all right to tell them about it. Then you should document that change.

Another point to consider when working with subcontractors is that your request for a proposal (RFP) can require that periodic reviews be included in your contract. This is desirable, as it would be if your customer required periodic reviews of your work. You are trying to see how their work is progressing and to understand if changes will be called for as a result of what they are doing or problems they are encountering; in short, you want to stay abreast of their work.

In many cases, you can do other things to help your subcontractors perform effectively. For instance, you may have one or more lengthy meetings with their project manager to be certain that he or she knows what is important to you. You can be willing to compromise noncritical items (with suitable contract changes, if appropriate). You can understand and review their schedule to try to offer constructive suggestions. You can check analyses and witness engineering tests, provided you that have made suitable arrangements.

However, you must draw a fine line between giving new directions and simply keeping abreast of what they are doing. Remember, the contract dictates the work. The progress reviews or monitoring activities are not a substitute for their management of their work; rather, these are **Your support agreement is a written contract.** solely for you to find out if the work is being done. You cannot send daily, weekly, or monthly changes in their direction and expect them to be successful.

TYPICAL PROBLEMS

Working with the support team probably causes the greatest difficulty, especially for new, inexperienced project managers. The root of this problem is being dependent on nonsubordinates. Two other problems are closely related. First, to negotiate support agreements takes a lot of time, usually at the very busy project inception period. Thus, it is done reluctantly or poorly or even omitted. In the latter situation, the project manager uses his or her own judgment of what the support group will do. Second, even when the support agreements have been intelligently negotiated, later events frequently require that changes be made. Again, this is time-consuming and must be anticipated. Here is a comment from a project manager one of us encountered:

(continued)

> I work on a project in which about 40 to 50 percent of the technical work is being done by another division. Both divisions' work is critical for successful completion, since the purpose of the project is to combine two technologies. The other division has serious technical difficulties and has been unable to produce a working device. The other division's team has resisted our technical suggestions, despite a looming project deadline. We believe our division can take over that portion of the project and do a better job.

This comment illustrates the challenge of working with and depending on internal support groups. [In fairness, there also may be issues of technical arrogance on the part of the division from which the comment was obtained, and there may be a not-invented-here (NIH) attitude on the part of the other division.]

Another problem arises when the support group people are extremely busy and already have a heavy work schedule. In this situation, it is often very difficult to obtain support for the project, and the project manager lacks any direct authority. A comment from a project manager illustrates another aspect of this problem:

> A project may require multiple departments or divisions. You get everyone to commit to the project at the proposal stage. After you [contract for] the project, it is not unusual for the other departments to put priority on their projects and not the tasks being performed for you.

Obviously, you should try to anticipate these potential problems when the initial plans are made. You also should beware of expecting effective help from anyone assigned to work on more than two or three projects simultaneously. Effectiveness declines significantly for such persons.

Finally, resist the temptation to recruit too large a team. Larger teams impose an inefficiency "tax" by requiring more time for communication. When two firms were developing comparable mainframe processors, the smaller team achieved a faster schedule.

HIGHLIGHTS

- The support team (or extended team) is a vital part of the project.
- Support groups should be involved in projects as early as possible and allowed to plan their tasks.
- You can improve communications when you formally document your assumptions and agreements.
- Network diagrams facilitate coordination of the activities of the support team.
- Distribute status reports and revisions to project baselines as established in your project communications plan.

15

The Role of the Project Manager

The project manager is ultimately responsible for making the project successful. Of the many competencies that the project manager must possess, perhaps the most important is that of leadership and influencing others.

The project manager is the "chief executive officer" of the project and must orchestrate a variety or resources to accomplish the project successfully. By definition, management is the project of getting work done, often through others. There can be little argument that leading a project involves engaging and motivating people.

PROJECT MANAGER COMPETENCIES

In recent years, there has been considerable research into those characteristics and skills that distinguish capable individuals. The following list describes selected characteristics that the profession considers project management competencies:

- *Business acumen.* Projects take place in an organizational and cultural context. Project managers must have some understanding of the purpose of the project and the "politics" of their organization.
- *Building customer relationships and stakeholder expectations.* As we emphasize throughout this book, customers and stakeholders have different ideas and metrics of project success. Competent project managers have the knowledge and skills to engage with others and to develop agreements with them.
- *Demonstrated knowledge of project management.* There are now a number of programs available to "certify" the knowledge of project managers. The best known of these is the Project Management Institute's PMP certification, but

there are others. Many companies now expect the PMP as an entry-level qualification to their own certification process.

- *Use of project management tools.* Numerous tools and software programs are available. Competent project managers have some skill in one or more of them.
- *Leadership.* Leadership involves many factors, including the presence of passion, empathy, inspiration, vision, and courage. We include the ability to influence others—independent of the formal organizational authority—as influencing and describe influencing in more detail in this chapter.

PROJECT MANAGEMENT CAREER PATH

Now that we have described a few project manager competencies, we want to extend the discussion to a formal career path for project managers. Over the last decade, many leading organizations have found the career path as one of the best ways to ensure good project performance.

> **One of the most effective things an organization can do to foster improved project management is to develop a career path to get into and advance in project management.**

As background, note that many companies have dual career tracks—technical and managerial—in the recognition that technical and managerial competencies are distinctive and of equal value. There is no sense in frustrating an outstanding technical person by "rewarding" him or her with a managerial position if he or she has no interest and talent in a management position. Organizations have found that they can better retain their talent by giving people career options and a path to follow that best fits their inherent skills and interests.

Here is a suggested a project management career path, which we developed by blending career ladders from several organizations. This summary provides some context for evaluating the basic-to-advanced responsibilities of individuals in the project and program manager role. Note that there is a steady progression from technical and tactical characteristics to strategic and integrative characteristics.

- *Individual project contributor and task leader.* This individual has knowledge of technical subject matter and is able to define and complete a work package, including coordinating the efforts of other individual contributors.
- *Project manager (subsystems level).* This individual demonstrates knowledge of all nine project management knowledge areas (which we described in Chapter 1). He or she can manage all elements of a less complex project, including customer interface, supplier interface, and team building.
- *Project manager (systems level).* This individual can perform all responsibilities of project management for relatively more complex projects. Such a person has a high degree of business acumen.

• *Program manager.* This individual can manage a portfolio of projects, providing direction to project managers of the individual projects. This individual demonstrates knowledge of program management and portfolio management.

Let's further explore the distinction between the individual performer and the project manager. Often, organizations will designate an individual the "project manager" because that person has been an excellent programmer, circuit designer, or carpenter rather than because he or she has been trained or has demonstrated competency as a project manager. However, a virtuoso technical performance is not a sufficient qualification for managing the efforts of the project and support teams. In fact, one's demonstrated technical or professional skills frequently are problem-solving or technical skills that do not involve an ability to interact with others. Developing people skills can be extremely difficult for many technically trained people who become project managers. Physical systems tend to behave in repeatable and predictable ways; people do not. In short, the role of project manager may be difficult for a specialist with a narrow "comfort zone." On the other hand, there are thousands of people who *do* add the necessary "soft" skills to their "hard" technical skills. John D. Rockefeller is reported to have said, "I will pay more for the ability to deal with people than any other ability under the sun."

Figure 15-1 distinguishes the skills used and time spent on different aspects of work by senior executives and individual workers. Project managers, in common

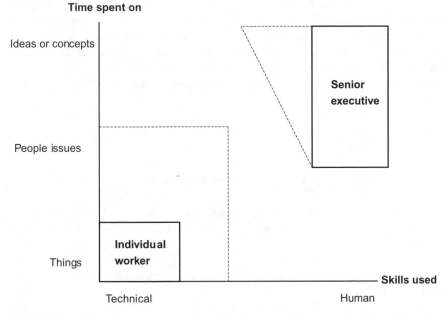

FIGURE 15-1. *The project manager must bridge two very different worlds. (Dashed lines indicate that skills used and time spent by a senior executive or individual worker often extend beyond depicted narrow worlds.)*

with functional line managers, normally occupy a middle region between these two very different worlds.

Project managers must be people-oriented with strong leadership and superb communication ability. They must be flexible, creative, imaginative, and adaptable to cope with a myriad of unexpected problems. They must be willing and able to take initiative with assertiveness and confidence in the face of substantial uncertainty and, in many cases, when confronted with significant interpersonal conflict. Finally, because projects are temporary, they must have a tolerance for the discomfort that comes with ending a project and finding a new assignment.

> **Integration expertise is as important as technical expertise.**

Finally, we are aware of some organizations' policies that state that demonstrated good performance in the project management role is required for promotion into executive ranks. There is considerable merit in comparing the similar responsibilities of a project manager and a chief executive officer. Both have to make decisions in complex environments and accomplish work—through others—with finite resources.

WHAT A PROJECT MANAGER DOES

Influence Rather than Authority

Throughout this book, we have emphasized that influencing is an important project management skill. While there is a time and a place to exert formal organizational authority by issuing commands, generally speaking, it is better to motivate people by aligning the needs of the project with their own individual interests and values. Project managers must operate by winning the respect of project and support team members. This accomplished, they will find that their wishes are carried out voluntarily and frequently with enthusiasm. One project manager highlights the challenge:

> How do you manage a project when you have overall responsibility but no authority over some of the work?

Another project manager states it somewhat differently:

> I do not have direct control over the external support groups. . . . How do I get the managers . . . of these . . . groups to complete the required tasks on or before the deadline?

A study by Thamhain and Wilemon identifies nine influence bases available to project managers:

1. *Authority*—the legitimate hierarchical right to issue orders
2. *Assignment*—the project manager's perceived ability to influence a worker's later work assignments

3. *Budget*—the project manager's perceived ability to authorize others' use of discretionary funds
4. *Promotion*—the project manager's perceived ability to improve a worker's position
5. *Money*—the project manager's perceived ability to increase a worker's monetary remuneration
6. *Penalty*—the project manager's perceived ability to dispense or cause punishment
7. *Work challenge*—an intrinsic motivational factor capitalizing on a worker's enjoyment of doing a particular task
8. *Expertise*—special knowledge the project manager possesses and others deem important
9. *Friendship*—friendly personal relationships between the project manager and others

The first of these nine elements depends on higher management's decision to delegate formal authority to the project manager and other individuals. Many people working on a project may not report directly to the project manager (e.g., the support team we defined in Chapter 13), and he or she does not have complete control over even those who do. In the first place, people are free to change jobs in modern society. If given a command they do not like, some workers will ignore it or quit the job completely or transfer to another division of the organization. The next five influence bases may or may not be truly inherent to the project manager's position; others' perceptions are most important in establishing their utility to the project manager. The seventh is an available tool anyone may use to influence others. The project manager must earn the last two. Projects are more likely to fail when the project manager relies on authority, money, or penalty to influence people; success is correlated with the use of work challenge and expertise to influence people. Leadership is exhibited by a person who is followed by others in the conduct of the undertaking.

There will be occasions when the project manager must negotiate with team members. A typical approach is to explain the rationale of the effort and to involve the people in planning the detailed work packages. Given this need to influence, an effective project manager must be a superb communicator. He or she must have verbal and written fluency and be persuasive to be effective; the next chapter contains some practical tips on how to improve your communication skills.

> **The project manager should focus on developing and applying influencing skills.**

Effective Managerial Behavior

As we wrote in Chapter 13, the project manager must work with people not of his or her own choosing, many of whom have different skills and interests. Furthermore, the project manager is a manager, not a doer. If the project manager is writing lines of computer code for a programming project, designing a circuit for a new product project, or nailing boards together for a construction project, who is planning the work of others? Who is deciding what approach to take to the support group manager so as to obtain the services of the most senior and best-qualified person? In addition, who is trying to devise a contingency plan in case the system test does not produce desirable results? The project manager must spend his or her time working with people and planning their work so that nothing is overlooked and contingency plans are ready if needed.

A manager must plan and manage. On a very small project, the project manager's participation is also required as a worker, not merely as a manager. If not physically, then at least mentally, a project manager in this situation should have two hats, one labeled "project manager" and the other labeled "worker." As we wrote in Chapter 1, the project manager must realize which function he or she is performing at any given moment and wear the appropriate hat.

Working with People

Occasionally, people find the need to issue "commands" and apply formal organizational authority. If stated brutally or insensitively, they demotivate and create resentment. Elias Porter has shown that we behave differently when everything is going well and when we face opposition or conflict. People also differ from each other. Some have an altruistic-nurturing orientation, others have an assertive-directing orientation, and others have an analytic-autonomizing orientation. Although altruistic-nurturing-oriented people usually are trusting (a strength), people of another orientation may see them as gullible (a weakness). Similarly, the assertive-directing person's self-confidence (a strength) can be seen as arrogance (a weakness), and the analytic-autonomizing person's caution (a strength) may appear to be suspicion (a weakness). A person with a balance of these orientations, who is flexible, may be seen as inconsistent. These strength and orientation differences must be understood if you wish to deal effectively with varied people.

In the case of complex projects such as new product development projects, the variety of required professional skills is exceptionally large. Such projects may involve marketing, research and development scientists, engineers of all sorts, software system analysts and programmers, industrial stylists and designers, procurement, manufacturing, finance personnel, and many others. Unfortunately, the training for each of these fields tends to be narrow, with the result that a specialist in one discipline frequently lacks the knowledge and language fluency to understand and collaborate with people from another discipline. One vital role of the project manager is to ensure effective communications among these diverse groups.

The project manager should have an interest and skill in human relations. To compound this problem, the same thing said to the same person at two different times can produce different reactions. This lack of predictability can be a major pitfall for many prospective project managers. Project managers must deal with both technical and emotional issues. If not already fluent in these human relations skills, they should seek training or coaching.

⚒ The project manager sets objectives and establishes plans, organizes, staffs, sets up controls, issues directives, spends time working with widely varied people, and generally sees that the project is completed in a satisfactory way, on time, and within budget. The project manager does

Managers manage, and workers perform the tasks.

not do the work of others on the project. A project manager who is an excellent electronics engineer may find it frustrating to watch a junior engineer carry out the circuit design activities on the project. The junior engineer will take longer, make mistakes, and not do as good a job as could a project manager with that technical skill. If the project manager starts to do the circuit design, though, it demotivates the junior engineer and lessens the manager's time to function in the most vital role of all, namely, that of project manager.

THEORIES OF MOTIVATION AND THEIR IMPLICATIONS

Regardless of hierarchical authority, anyone can encourage or stimulate others to contribute to a project and improve their productivity. For project managers, many of whom lack direct authority over all the resources required for project success, an understanding of motivation is essential. Here are two comments from project managers illustrating the challenge:

How can I motivate these individuals?

A project . . . needs the involvement of several other employees outside of the original team that was chosen for the project. What is the best way to not only include more employees in the project but motivate them to do work outside of what is required of them?

As you will see in what follows, a major element of the project manager role is to avoid demotivating others.

Consider the following problem, which a project manager called to the attention of one of us (MDR):

Motivation is internal and based on perception of needs.

If I fail to meet a schedule I prepared myself, I feel obliged to work overtime (frequently unpaid) to try to meet the estimate; however, if the schedule or estimate was prepared by someone else, 1 generally feel under no obligation to try to meet it. That is, the work takes whatever it takes. I'm not sure this is a reasonable attitude. Nevertheless, regardless of what is logical, it is the way I actually feel that controls my motivation.

Another project manager described the situation of a coworker who had been severely criticized by his boss. As a result, that coworker is no longer motivated and is both argumentative and defensive about many things he is now asked to do. In both these situations, motivation—or the lack of it—is a challenge for the project manager.

An individual's feelings may appear illogical to others, but these govern his or her willingness and enthusiasm to carry out work on your project.

There are many theories of motivation, four of which we briefly summarize and provide discussion of the project management implications. Thus, an understanding of motivational theory is important. Although the basic work is decades old, it remains important for you to understand. By analogy, we still teach Euclidian geometry and Einsteinian relativity even though they were formulated a long time ago.

Hierarchy of Needs

People are motivated to fill their unsatisfied needs, not their manager's unsatisfied needs.

Abraham Maslow's theory of the hierarchy of needs (Figure 15-2) describes (and human experience validates) that motivation is internal to a person's perceptions of five levels of needs. The worker is motivated to achieve a specific goal because of an inherent internal need. The first of these is physiological or body needs (i.e., eat, sleep, have shelter, and so forth). A hungry person will have a goal to eat and will engage in the goal-directed activity of buying and preparing food to satisfy the hunger need. (In fact, if we try to prevent this person from satisfying the need, by denying him or her money to pay for the food, for instance, the person may engage in such antisocial behavior as robbery to satisfy the need.)

To influence people, you need to address them at the level of unfilled need they have.

Once this need is filled, continuing to offer more food, sleep, or shelter has no motivational value. Higher-level needs now come into play. Second-level needs are safety and security, third-level needs are social, and fourth-level needs are esteem or ego. The fifth-level needs are for self-actualization or fulfillment. (Recent motivational theory refinement proposes two additional levels between Maslow's

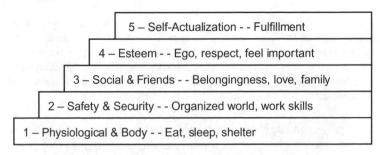

FIGURE 15-2. *Motivation—Abraham Maslow's hierarchy of needs.*

fourth and fifth levels: cognitive, the need for knowledge and understanding, and aesthetic, the need for order and beauty. These additions do not alter the basic lesson for project managers.) To take advantage of Maslow's findings, a project manager would have to understand the levels of need an individual has already satisfied. Then he or she could offer satisfiers for unmet needs as an encouragement. Americans generally have unfilled needs on levels three, four, or five; therefore, the project manager would offer satisfiers aimed at these levels.

If the project manager does not offer satisfiers, workers will find them outside the work environment (Table 15-1). Note that money is not a motivator, although it may play a role. A person feels good when he or she gets a salary increase because it provides recognition (filling an ego need). However, if a person can fill social needs only by joining an expensive country club, he or she needs more money than the person who can satisfy those social needs with informal activities.

Finally, you must realize that there are some tasks on a project for which it may be impossible or exceptionally difficult to motivate a worker. One project manager asked one of us, "How do you motivate people to follow through on the less glamorous aspects of the project, such as punch list items [i.e., minor residual details]?" He could have asked about the necessity to clean up laboratory equipment or other similar tasks. The best suggestion is to try to make people realize that all work is a mix of the mundane and the exciting and that they should harvest some of the euphoric "high" they get from the stimulating tasks to carry them through the depressing "low" of the necessary but boring ones.

Motivational Factors

Frederick Herzberg has done an excellent study on work and motivation, examining specific factors that motivate workers. He found many things done (e.g., company policies, supervision, work conditions, and salary) that are not substantially motivational. He called these "hygiene factors." The absence of hygiene factors is demotivating, but their presence is not motivating.

TABLE 15-1 Examples of Satisfiers of Unfilled Needs in Maslow's Hierarchy

Maslow Level	Job-Related	Other
1	Sufficient salary and good working conditions	Availability of desired stores and services
2	Company training programs	Comfortable home and lifestyle
3	Company-sponsored team building	Close friends and a social network
4	Acknowledgment of productivity and promotions	Leadership positions in social, charitable, or religious groups
5	Choosing own projects	Achieving personal goals

Achievement and recognition are powerful motivators.

Instead, there have to be motivational factors present to foster high productivity. The key motivational factors are achievement, recognition, the work itself, responsibility, advancement, and growth. Achievement and recognition are short term, and the others are long term in their impact. Thus, it is important to give workers recognition frequently (but not routinely) for significant accomplishments.

Theory X/Theory Y

Douglas McGregor's "Theory X and Theory Y" work posits two sets of assumptions that people make about others. Managers with a Theory X assumption assume that people lack desire to do a good job and to even work. Consequently, a manager who holds a Theory X mind-set believes that incentives or disincentives will cause action on the part of the worker. The mind-set often manifests itself in an authoritarian style, where the manager attempts to coerce the other person to comply.

Participative decision making has many advantages.

Theory Y is based on the assumption that others want to do a good job—that people both enjoy working and want to work. The mind-set often manifests itself in the participative style; managers can draw on workers' self-direction. The job of the manager becomes more focused on communicating requirements and removing barriers so that the individual can do a good job.

While the real world of projects does include psychopaths and saboteurs, we find that most people sincerely want to do a good job on their projects. Thus, we admit to a bias to Theory Y. Still, this does not mean that tactics that stem from Theory X assumptions are irrelevant or useless. Sometimes people do need incentives. Sometimes people *do* need a "kick in the pants." Sometimes the only way to get people to focus on what you want them to focus on is to insist on delivery by a due date.

Behavior Modification

B. F. Skinner devised a theory of behavior modification that advocates positive reinforcement, namely, rewards for "good" behavior. To apply Skinner's theory, one would induce people to behave differently (i.e., consonant with project goals) by rewarding them when they act appropriately. "Rewards" typically are consistent with Herzberg's findings, namely, a sense of achievement and recognition.

Negative reinforcement can stop undesired behavior, but it requires positive reinforcement to promote desired behavior.

Thus, if a designer does a fine job on your project, it is appropriate to send her a memo and send a copy to her boss (or vice versa) and perhaps to her personnel folder as well. Conversely, if you have been practicing behavior modification and positive reinforcement consistently, you do not have to do anything if on another occasion she does a poor job. The absence of positive reinforcement will be message enough. Further, that absence may very well motivate the designer to ask you how her performance fell short.

Very early in the industrial career of one of us (MDR), the president of his company sent the following letter (on engraved personal stationery) to him at his home just after the project team of which he was a member shipped the first unit of an advanced system:

> Dear Milt:
>
> I would like to extend to you my hearty congratulations for your contribution to the outstanding technical success of the Satrack Program. I know that you must feel proud of being a member of the team who accomplished a marked advance in the state of the art of aspheric manufacture.
>
> Our Company is now considered to be in the forefront of this development activity, and it has been through your contributions that we have achieved this position.
>
> I know that your efforts were great and there were many long evenings and weekends which you personally sacrificed. It is indeed gratifying that we have the people with the spirit to undertake such a challenging problem and carry forth to a successful conclusion.
>
> Sincerely yours,

The recipient of such a letter will become motivated to put in similar extraordinary efforts in the future and probably will continue to have family support for that effort. Other examples of positive reinforcement include awards (e.g., wall plaques, luncheons or dinners, trips, and so forth) or a story or picture in the company newspaper. These promote worker motivation because they provide recognition and validate an achievement.

People's behavior can be modified.

THREE USEFUL TECHNIQUES

Management by Objectives

Another practical technique that a project manager can use is *management by objectives* (MBO). Variations of this technique are *management by results* (MBR) and *management by commitment* (MBC). As the names suggest, the techniques focus management and worker attention on the outcome (objectives or results) rather than on the process by which a worker chooses to achieve the outcome. This frees the manager to concentrate on what is desired, and it leaves the worker free to concentrate on how to accomplish it in his or her own style.

To use this simple, powerful technique, the project manager negotiates with a worker on the results the worker will agree to achieve. Because the technique is a general one, the worker may be a member of the project team or the support team.

The agreed results must be like project objectives, that is, verifiable, measurable, specific, and achievable. These are recorded on paper (perhaps a standard form) and are signed by both the worker and the manager. In the event that changes are required, these must be jointly negotiated, and the paper then must be revised and signed again. Note that this technique is identical to how you would contract with another organization to do subcontract work.

The technique draws on the motivational factors of responsibility and achievement. The worker may be motivated because he or she is consulted in a responsible way in defining the task to be carried out. Moreover, there may be clear progress during the work that provides the worker with a periodic sense of achievement.

The only problem with the technique is that it is sometimes difficult to set the objectives or results that are to be achieved. When this is the case, the same technique used for starting a project with unclear objectives can be used. That is, agree on something specific that will be accomplished in a given period, and then try to set longer-term goals when more is known about the task. Conversely, if you cannot establish a specific goal for a worker to achieve, it is clearly unfair to hold him or her accountable for something you could not describe.

Establish a Creative Environment

Frequently, the project manager also must stimulate creativity. This is required to some extent on all projects, but it is especially applicable to high-technology projects. Even two similar projects, for instance, building two identical radio telescopes, can encounter different soil conditions requiring an innovative approach in one case. In general, the less precedent for the project, the more creativity will be required.

Creativity may be stimulated simply by managing in a way that is consistent with behavioral and motivational theory. Encouragement by providing recognition and appreciation is the most straightforward technique to stimulate creativity. In addition, one must provide a favorable atmosphere. In a sense, we are looking at positive reinforcement. People are permitted to fail when asked to produce creative results and are not castigated for doing so. Rather, they are praised when they succeed.

Brainstorming

Positive reinforcement and brainstorming will stimulate creativity.

Brainstorming techniques are used often to deal with some intractable problems. The conventional method for brainstorming is to advise perhaps a half-dozen people of the problem and after one or two days convene a brainstorm meeting. At this meeting, restate the problem and reiterate the ground rules:

1. Absolutely no criticism (including smirking) is permitted.
2. The more ideas produced, the better is the session.
3. Novel, unusual (even impractical) ideas are desired.

4. Improvement or combination of prior ideas is also desirable.

Use a tape recorder to permit more leisurely subsequent consideration of the ideas thus generated.

An alternative method that works better in most situations is to have a facilitator talk to a few people individually and ask for their ideas on solving the problem. After three or four people have been interviewed, the facilitator will have a list of ideas to use to start the brainstorming session *Facilitated brainstorming is very productive.* (like pump priming), which is then carried out in the normal way. This facilitated brainstorming produces more ideas and they are of better quality.

TYPICAL PROBLEMS

In many cases, project managers are not selected from a pool of trained, qualified people. (This is particularly true when a company does not provide a project manager career path.) Rather, projects arise within (or descend on) the organization, and a person who has demonstrated technical proficiency is asked to become project leader.

Such people are often good "doers" and have technical skills and may think that they want to be a project manager, but they usually take the job not knowing what is involved. In general, technical specialists rapidly master planning techniques and then the mechanics of project monitoring. They may get along with others (as opposed to being hermitlike) but be unable to cope with the inevitable conflicts that bedevil the project manager. Or they may be poor communicators. What then happens is the organization has a poor project manager and has lost the services of a good technical specialist.

One cure for this problem is to be sure that candidates for project management read books such as this prior to being offered jobs as project managers. After that, assuming a continuing interest in the job, the selected candidates should be offered further training, as discussed in Chapter 27.

HIGHLIGHTS

- Good project managers have many competencies.
- Consider the career path into and out of project management.
- Project managers should learn how to wield influence.
- Human relations skills are vital to a project manager.
- Familiarity with theories of motivation will help managers to do their job.
- Creativity can be stimulated by positive reinforcement and brainstorming.

16

Practical Tips for Project Managers

Because a project manager must be a superb communicator, we first discuss the general problem of communication and then provide several simple suggestions for improving your communication skills. Another pervasive problem for project managers is resolving conflicts, and some techniques to deal with conflict are provided. We also briefly discuss how a manager can gain time to work more effectively with people.

COMMUNICATION

No matter how hard individuals try to make themselves understood to others, communication is difficult. There are so many obstacles that it is amazing that any

It's not what you say that matters; it's what they understand.

effective communication occurs at all. Words have different meanings, and people often have different perceptions or orientations. The project manager's reputation (be it as a jokester or as a very serious person) will alter the way any message is received. Everyone the project manager communicates with will tend to hear the message he or she

wants or expects to hear, which is not necessarily the message the project manager is attempting to deliver. Sometimes people are not listening, are distracted, or have a closed mind.

Communication involves a sender, a message, a medium, the receipt and understanding of the message, and feedback to the sender.

Communication must be worked at.

There is an aphorism about how to communicate: "First, you tell people what you intend to tell them; then you tell them; and then you tell them you told them." There is much truth in this use of multiple message delivery

There are several general steps you can take to improve your communication with other people:

- Just as you should have an agenda before a meeting, you should plan what is to be communicated beforehand rather than trying to decide while communicating. As it is sometimes stated, "Put brain in gear before opening mouth."
- Use face-to-face meetings in which you can observe the other person's "body language." Allow enough time at an appropriate time of the day.
- Decide which sequence and combination of telephone discussion, face-to-face meeting, and memo will be most effective.
- Use e-mail carefully and thoughtfully. Much e-mail is quite casual, is sent hastily or carelessly, is given insufficient attention by recipients, and can cause unnecessary confusion. E-mail does have the advantage of quickly reaching many people at dispersed locations if they are connected online. Of course, they must have time to read, digest, and act on this torrent of messages.
- Be consistent and follow through with actions appropriate to your message.
- Use simple language.

Feedback

Communication is very much like a servomechanism in that it is not effective unless there is feedback. Communication can be improved by asking the person to whom the message has been delivered to restate it in his or her own words. This can help to overcome a listener's closed mind. Another effective technique is to back up any verbal communication with a memo. This also may be done the other way around, first sending the memo and then having a meeting to discuss it. The duality of mode and the recipient's restatement, rather than being simple redundancy, is most effective here.

Feedback improves the likelihood of effective communication.

Notices

It is impractical to meet constantly with all participants on a very large project. Even on a smaller project, it may be disruptive to have numerous meetings. It is thus desirable to issue project notices and reminders of priority actions for any given period. If and when you transmit information by paper, consider using distinctively colored paper or preprinting the project name on the top to set the notice apart from other mail. In addition, e-mail and other forms of modern telecommunications can be used. This is why many project managers find that e-mail is their most widely used project management tool.

Proximity

Locating the people on the project near each other also aids communication, as we discussed in our chapters on team building. Because the people are close together, they can see each other more often, which makes communication easier and more frequent. When people are in frequent contact, their points of view tend to become more uniform.

Follow-Up

It is necessary to have some system of follow-up of the communications, be they electronic, face to face, or written. Some people simply keep an action log, a chronological listing of all agreements reached with other people for which follow-up action is expected.

Somewhat more effective is a follow-up system keyed to the individual from whom action is expected. A filing card with each key person's name printed on the top may be used to record notations of actions expected of that person. A variant of this is keeping a folder for each key person in which you store records of all discussions or copies of memos for which follow-up action is required or requested. In either event, hold periodic meetings with each key person, and use the filing card or folder to plan the topics to be discussed.

Follow up communications.
When it is known that project managers (or any manager, for that matter) have such a consistent follow-up system, people who work for or with them will realize that any statements made to them will be taken seriously. Therefore, commitments made to them tend to receive serious and consistent attention.

CONFLICT RESOLUTION

Projects are fraught with conflicts. They inevitably arise because projects are temporary entities within more permanent organizations. One root cause is competition for resources. Another cause is illustrated by the following comment from a project manager:

> The project required that we adopt a standard style for the drawing package. However, the instrumentation and electrical engineers on the team had incompatible personalities and work backgrounds. This produced ongoing conflict.

The project manager must manage conflict.
Regardless of organizational form, project managers and functional managers tend to have momentary interests that are at odds, so project managers must expect and be able to "stomach" conflict. If you have a low tolerance for conflict, being a project manager can be frustrating.

Since finite resources are normal, and there are many differences of opinion on who needs the resource, it is worthwhile to consider these three additional conflict-resolution strategies:

- *Arbitration.* This method involves a third party who listens to the arguments of each party and makes a decision on which party will prevail in the conflict. It is often a preferred method when two organizations have a legal disagreement, but it is also the process used when a program manager or senior manager decides that "Project A" will get a scarce resource that "Project B" also desires.
- *Mediation.* This method involves a third party who "splits the difference." It is a way that allows both parties to a disagreement to feel that they have been treated equally well (or equally poorly). For example, a program manager or other senior manager would decide that "Project A" and "Project B" each would get 50 percent of the available resource and accept the consequences of the changed estimating assumptions.
- *Negotiation.* This method involves give and take between the parties to a disagreement. For example, the project managers of "Project A" and "Project B" would meet and compare their resource baselines and their effect on the project critical path, making agreements between themselves to reallocate resources between their projects.

Consider these three ways to proactively reduce conflict:

1. Work at conflict reduction proactively because conflict is always present and will not go away if you ignore it or wish it were not present. In fact, left unconfronted, it normally gets worse.

   Constantly anticipate and try to reduce conflict.

2. Have and maintain good plans in the form of current and realistic schedules agreed to by the people involved.
3. Communicate thoroughly with all people involved and their management.

EFFICIENT TIME MANAGEMENT

Given the wide range of project managers' duties (in a sense, they must be all things to all people), they can easily end up working nights and weekends unless they are very efficient in the use of time. Of course, they should not make the mistake of being efficient to the point of being ineffective. Effectiveness is achieving the desired results. Effectiveness is what counts, but project managers are more likely to be effective if they use their time efficiently.

Manage time efficiently.

The overriding issue in time management is "first things first." The project manager must know the most important things to do this year, this month, this week, today, and right now. One project manager told us that she spent two-thirds of her available time in meetings, focus group testing panels, training seminars, and serving on committees and that only about one-eighth of that (two-thirds) time had anything to do with the new product development projects for which she was responsible. Only when he or she has a clear perception of priorities can a project manager manage his or her time effectively. Thus, each day select a few (not more than three or four) really important, high-impact matters for your primary attention.

The second key tool in effective time management is to devote large chunks of time to important single issues (e.g., one two-hour meeting with a colleague rather than twelve fifteen-minute meetings). This can be accomplished by maintaining a time log on how you actually spend your time during a week. A time log is a simple grid with time intervals of 10, 15, or 30 minutes down the left edge and five columns, one for each day, across the top. Maintain such a log throughout the week you choose to sample. At the end of the week, examine the record of what actually occurred, as you recorded it at the end of each time interval during the week. Then plan how you will alter your behavior in the following week to get fewer, larger chunks of time concentrated on single topics. Do this several times over a period of several months, and you will master the art of better time management.

Because project management involves integrating the work of many people, numerous meetings will be held. As some anonymous sage has stated, "Meetings are an institution where the minutes are kept and the hours are thrown away." Conducting them efficiently and effectively is essential. The following are keys to improving meetings:

- Know beforehand why the meeting is to be held and what outcome is expected. Consider whether it is possible to omit the meeting.
- Determine the minimum number of people required.
- Choose a meeting location with a room arrangement consonant with the meeting's purpose (e.g., a round table arrangement for discussion among equals, a lecture hall arrangement for a presentation, and so forth).
- Circulate an agenda with topic durations to all attendees, and perhaps discuss this individually with the key participants ahead of time.
- Be prepared, and open the meeting on time with a restatement of the purpose and agenda.
- If possible, ask each attendee (one at a time) for his or her views on each topic prior to topic completion. Just because some attendees are loud or dominant does not mean that they are best qualified to speak on a topic; a quiet, shy, or retiring person often will make a valuable contribution if invited to comment.
- Verbally summarize what transpires at the meeting, and later distribute published minutes to all attendees.

Plan all meetings, and always circulate an agenda in advance.

They won't work everywhere, but you might want to try the meeting rules on the Gossamer Albatross project (as reported in *Technology Review,* April 1981, p. 56):

- All meetings are held standing in a circle.
- All participants are heard in turn.
- All meetings must result in a definite decision.
- All decisions must be acted on immediately.

A 1999 University of Missouri experiment found that seated meetings lasted 34 percent longer than standing meetings, but the quality of decision making was no greater (as reported in The *New York Times,* June 22, 1999, p. Y D7).

TIPS

There are a few "tricks of the trade" that project managers should employ to help put the techniques discussed in this chapter to practical use.

First, keep your "door" open. While fewer managers these days actually have offices where doors can be closed, they send out signals that they are disinterested in the concerns of others. Keeping your door open encourages people to talk with you, and sometimes that will identify key project issues that you were not yet aware of.

Second, close your "door" and do not answer your telephone, e-mail, pager, or allow other kinds of interruptions. This is when you do your planning, to gain high leverage on time use. When you are also a worker on your project, which is a common situation on smaller projects, this is when you do your own project work. In addition, a closed door permits private discussions if these are required. Finally, a closed door is a trivia filter, which may force people to stand on their own feet and make minor decisions.

Third, "walk the halls" if people are located at the same facility as you. There are always some people who will not enter your office, even if the door is open. Also, what you inevitably see when you go to the sites where work is supposed to be done is that things are not as you expected them to be. At Hewlett-Packard, this is called *management by walking around* (MBWA). If team members are not lo-cated at the same facility as you, get out to see them, or at least phone them.

Fourth, set a good example. Arrive early (or at least punctually) for work, take your job seriously, work hard, and be respectful of others (especially upper man-agement and your customer), even if you disagree with some of their actions.

Fifth, when you are encountering problems, do not try to conceal them. Rather, seek advice from more senior people.

Sixth, remember that one of the most effective things a project manager can do is change resource allocations. Rather than pushing harder on (or shouting at or abandoning) a task manager who is experiencing difficulty, try to get him or her additional help.

TYPICAL PROBLEMS

The development of your own natural style of management may be a problem when applying the practical tips in this chapter. Our styles of dealing with conflict or improving time management, for instance, may not be best for you. You must find your own way to deal with limited time and multiple demands on it, and you must recognize that interpersonal conflicts can be very distracting and impede productivity.

Another problem is balancing the two roles of project manager and worker if the project is such that you must play both roles. Consider this problem cited by a project manager:

> I am in charge of a project that involves a number of off-site consultants. Several of these consultants are very busy and are difficult to reach. It is nearly impossible to get everyone available at the same time to hold meetings. How can I most effectively exchange information with these people without spending too much of my time on the phone instead of working on the project?

The key point this problem illustrates is that the project manager can delegate everything other than the managerial role, so he or she must spend less time "working on the project" and make more time available to manage it.

In general, however, the biggest problem is to become a superb, versatile communicator. You must work with varied people—many (or most or even all) not of your own choosing—and that can be done only if you communicate effectively with them. Some of the people you must deal with may themselves be poor communicators, as either message deliverers or receivers. Some may receive and reply to written memos, and perhaps you can write these well. On the other hand (and this will always be the case with some workers), they may not read or understand written memos. Therefore, you must talk with and get useful responses from these other workers. Obviously, the reverse may be true also; that is, your writing skill may be relatively weaker than your speaking skill. Nevertheless, some workers receive written information better than spoken, so you may have to improve your writing (or reading) skills.

You must be sufficiently versatile to deliver messages that can be understood by varied people who have different preferred ways to receive your communications.

HIGHLIGHTS

- Effective communication is aided by feedback, issuing notices, and locating workers near each other.
- Conflict between people (or their group managers) must be expected on any project. Conflict can be reduced by having plans that are current and by good communications.
- A project manager can be more effective if he or she is a good time manager.

Part 4

Controlling the Project

17

Essentials of Project Control

Control is a fundamental general management function. This chapter describes the fundamentals of control as applied to projects. This includes monitoring of status, evaluation, forecasting, communication, and corrective action. We also discuss the program management activity of monitoring several projects simultaneously.

In our Chapter 2 discussion, we presented an analogy comparing a project to a plane flight from Los Angeles to Honolulu. The flight crew first files a flight plan specifying the planned course. In this analogy, the flight plan is the project baseline because it establishes a series of comparison points for project control. During the flight, the crew measures the plane's position, noting and reporting variances. The pilot assesses the variance and adjusts with corrective actions to ensure that the plane gets to its destination at the scheduled time. The purpose of control is to measure progress toward objectives, evaluate the extent to which anything needs to be done to reach the objectives, and then take corrective actions as necessary. Thus, you control your project to keep it on course. Control is an important general management function that we summarize with this acronym: BMAC, which many people remember as "big MAC." The acronym stands for *baseline, measure, assess, corrective action.*

Monitor actual progress, and compare performance with the project plan.

In this chapter, we described project control as an application of the following five practices:

- Develop a baseline schedule of work, resources, expenses, and so on.
- Develop a performance measurement system, including a communications plan.

- Measure performance against baseline, and determine variances as values, not percentages.
- Develop a forecast by first evaluating the status and then considering other factors.
- Take corrective actions on the variances.

DEVELOP A BASELINE

The integrated project plan is a model of how the project team intends to accomplish the project objectives, as we have described in earlier chapters. An important outcome of planning is the decision to baseline the project, and this typically includes a work breakdown structure (WBS), a network diagram that indicates every element of the WBS, and a cost estimate for each activity. A common weakness is that project teams do not develop appropriately detailed baselines for measuring schedule, scope, resources, risk, and similar project aspects.

People frequently blame estimating for the observed variances during execution. While it is true that estimating is difficult, as the say, "that is history." Once you have made a commitment to a baseline, you need to do the best you can to manage to the baseline. If you need to change the baseline plan, carefully read Chapter 20.

People will pay attention to what is measured.

Project planning and a documented project plan are an important foundation for project control. A good project plan is a resource for the team to self-manage its work. A deviation from the baseline indicates the need for corrective action.

DEVELOP A PERFORMANCE MEASUREMENT SYSTEM

As part of planning for the project control function, the project team should determine the reporting frequency and method of reporting. Answer the question, "Who needs to know what?" Reports can be summaries or detailed about work task and other elements. If the project organization is an industrial company, there are many people that might request a project report, as shown in Figure 17-1.

The first 15 percent of the project is critical, and you need to establish the monitoring early. You will be reporting status data on a number of elements, including the following: quantity of work accomplished, resources expended, and their variance compared with the baseline.

Because many people concerned with the project will wish to receive reports, there is a tendency to try to circulate a single report to many recipients. This is a mistake because top management will want summary-level status and forecasts, whereas middle managers will want more specific and tailored information on operational details. Reports always should be concise and consider the needs of the audience.

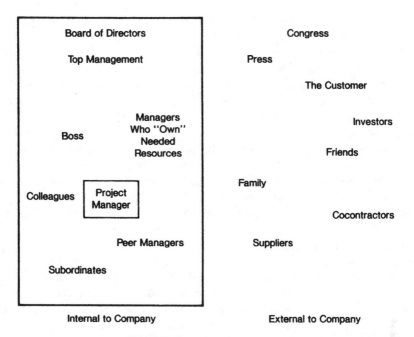

FIGURE 17-1. *Report recipients.*

Pictures, demonstrations, and models also should be encouraged, especially for a high-technology project or one where much of the work is geographically remote. It is often hard for people not intimately concerned with a project to visualize the status, expected outcome, or even the concept. For them, tangible descriptions, pictures, and such are by far the most appropriate means for providing reports. Consider using your organization's fax, video, and Web-conferencing capabilities to permit geographically remote sites to construct a model or visualize design information in a three-dimensional display.

MEASURE PERFORMANCE AGAINST BASELINE AND DETERMINE VARIANCES

Project status is a "snapshot" of "where you are now" compared with "where you planned." Status variation is the gap between actual performance and baseline. It is reported to an "as of" reporting date—not the entire project.

Status is best reported as a positive or negative variance. For example, negative variances of −1 week and −$15,000 dollars would mean that the project is behind its baseline schedule by 1 week and has spent $15,000 more than planned "as of" the reporting date. Avoid using "percent

Determine variances as values, not percentages.

complete" generalized for the entire project; for example, saying, "the project is 60 percent complete" really doesn't tell you much.

Good project performance management systems are based first and foremost on the work performed, not the money or time consumed. When a task is simple, such as drilling a hundred holes in a steel plate, it is easier to measure what percentage of the task is complete. (Moreover, even in this case, there is no assurance the last hole won't destroy the plate, forcing the task to be completely redone.)

Avoid guesses of percent complete.

Since so much of modern projects is intellectual, it is tough to actually measure the work. If you must accept a guess of work completed, document and validate a set of assumptions to support the guess. Why? People are tremendously overconfident in their belief in the accuracy of numbers that they generate as estimates. Generally speaking, people overstate their percent complete by about one-third: If they say that they are 50 percent done, likely they are only 20 percent done with the task.

Concentrate on completed task deliverables, not activities.

Consider the schedule situation illustrated in Figure 17-2. There is a long period of time in the middle of the project when there are no scheduled task completions. Thus, no certain checkpoints are available, leaving considerable uncertainty as to the actual status. This illustrates another reason to break a project into many small tasks, because there will be less uncertainty as to schedule status.

The most conservative approach to measuring status is the *0 to 100 percent criterion,* where only tasks for which all work has been done are considered complete. If the WBS divides the project into many small tasks, the manager can look at each of them individually and decide whether they are complete. Using the 0 to 100 percent rule, no other task, regardless of the amount of effort applied so far, is complete. A task that is reported as 80, 90, or 99 percent complete is not recognized as complete for status-reporting purposes. The project manager can accept as complete only tasks for which reports (oral or written) guarantee that the

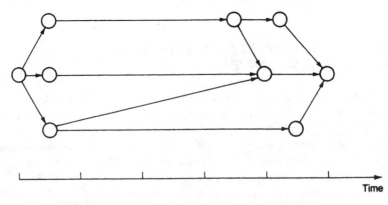

FIGURE 17-2. *Time-based activity-on-arrow diagram.*

activity is complete. This binary approach to crediting progress on tasks simplifies the project manager's job tremendously.

There are other approaches to measuring progress. For example, the *50 percent–50 percent criterion* would recognize 50 percent of the value when the task starts and 50 percent when it ends. Many project managers have used this approach successfully for work that involves intangibles. There are also other widely used measurement criteria such as the *20 percent–80 percent approach,* where you recognize 20 percent of the value when the task is started and the remaining 80 percent when the task is completed.

In a good project performance management system, report of percent complete should not be based on the amount of time or budget consumed (or an average of these).

The WBS tool is important to project control. The task authorization document (see Figure 6-4) could have additional blocks for the signatures of the task manager and project manager to indicate their agreement that the task is satisfactorily complete (Figure 17-3). With many small tasks rather than only a few larger ones, there is much less uncertainty about overall project status.

Dividing a project into many small tasks aids precise determination of its status.

Finding simple, highly visible measures can be helpful. Figure 17-4 shows one way to report schedule variance from plan. The number of days until completion is graphed periodically, and any deviation from the "no slippage" line (drawn at 45 degrees) reveals slippage. (A horizontal line shows no progress.) Figure 17-4 has a line above the "no slippage" line, indicating a small slip, which might be reasonable. Another simple device is to arrange for calendar clocks that count backward to the scheduled completion.

The project manager cannot depend entirely on documentation for reporting status. In the first place, documentation can lack accuracy for a host of reasons, for example, data entry errors or delays if a computer-based system is being employed. Further, the people who write reports are prone to unwarranted optimism. Project workers generally assume that a well-advanced task is nearly complete; in fact, no task is complete until it is truly finished. Thus, most task reports will indicate that a task is 80 or 90 percent complete, implying that it will require additionally only a small fraction of the time already spent.

Do not confuse activity with progress.

In Chapter 19, we describe the earned-value management system. Many organizations find that this is a powerful project control tool.

FORECASTS

Once that you have determined the current status (the variance between baseline and actuals), you can make your forecast about the actuals at completion. Recall

TASK AUTHORIZATION	PAGE		OF

TITLE			

PROJECT NO.	TASK NO.		DATE ISSUED

STATEMENT OF WORK:	COMPLETION REPORT:
APPLICABLE DOCUMENTS:	

SCHEDULE START DATE:	COMPLETION DATE:
COST: PLANNED =	ACTUAL =

ORIGINATED BY:	DATE:	ACCEPTED BY:	DATE:
APPROVED BY:	DATE:	APPROVED BY:	DATE:
APPROVED BY:	DATE:	APPROVED BY:	DATE:

COMPLETED BY TASK LEADER:	ACCEPTED BY PROJECT MANAGER:

FIGURE 17-3. *Task authorization form modified to serve also as a task completion report and acknowledgment form.*

Status first and forecast second.

that the status is a snapshot of how the project is progressing against the plan. The forecast is an estimate of the future based on your current progress and other factors such as outstanding risks or issues. Use the cumulative status as a basis for forecasting the percent complete for the project.

You should develop the forecast by first considering the cost and schedule variance. Is the variance a one-time event, or is it a systematic event? The one-time event is probably a fair assumption early in the project, and the systematic assumption is appropriate for the middle of the project.

For example, let's assume that you have an eight-month project with a budgeted cost of $80,000. Let's say that you are two months into the project and find that your status (as of the end of the second month) is that you have only completed one month's worth of work and you have an actual cost of $40,000.

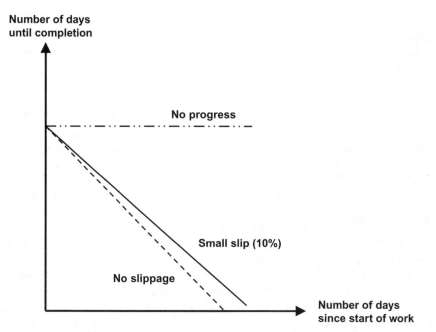

Number of days until completion

No progress

Small slip (10%)

No slippage

Number of days since start of work

FIGURE 17-4. A simple schedule report to quickly reveal slippage.

The cost variance is −$20,000, meaning that you have spent $20,000 more than you should have for the work installed. If this is a one-time overrun, you forecast completion cost as $100,000. If this is a systemic issue, then the forecast completion cost is $160,000.

You can apply a similar logic to schedule performance. If this was a one-time delay of the project, then the project is delayed by one month for a total project duration of nine months. However, if this is a systemic issue, then the forecast duration is 16 months. See Chapter 18 for other ways to deal with this issue.

Project performance management systems are important because they provide a guide for management action. They permit comparison of actual accomplishments in terms of cost incurred and planned accomplishments. Thus, they aid in determining whether there will be a cost overrun or underrun at completion, as well as forecasts of completion dates.

Now, it is up to the project management team to contemplate correction actions. For this example, if time is a more important element, the project can consider "crashing" the schedule by adding more resources, in effect, trading off time performance for cost performance. We will explore corrective actions further in the next section.

CORRECTIVE ACTIONS

Corrective actions could include withholding resources or discretionary authority. When the project manager uses these actions, he or she is assured that people working on the project request the use of these resources or authorities, thus providing visibility. As an example, the project manager could require any expenditure in excess of $1,000 to receive his or her specific approval. Alternatively, the project manager could require that any drawing release have his or her signature. These kinds of controls go beyond the project plan in that they make project workers seek out the project manager for approval during the performance of each project activity or task. Anyone's failure to request approval of a planned major purchase signals the project manager that the task potentially has deviated from the plan.

Controls can tell you if the project plan in being followed.

For the preceding example, not all decisions on a large project must flow through the manager; the project will get bogged down by his or her lack of time to review a myriad of documents for approval. A variation on this approach is to insist on independent inspection and quality-control approvals or on test data as a means to verify progress. For instance, you could insist that each subsystem test be approved by people working on other related subsystems.

Another method is to trust the intentions and actions of the person carrying out a particular task. This is a form of management by exception, and what you want to have brought to your attention is any "political" problem or deviation from performance, schedule, or budget. This method is fine if that person is able to do three things: (1) recognize deviations from plan, (2) report the deviations promptly to the project manager, and (3) report the problem clearly. Because these three preconditions are rarely satisfied, this control tool normally should not be used.

Control is a team activity.

A far better approach is for the project manager to examine the work being done under the direct control of the project team and support teams. This kind of control is based on the Theory Y assumption (discussed in Chapter 15) that people working on project tasks are trying to do a good job (which often can become a self-fulfilling prophecy). These examinations of activity often may be accomplished by reading reports and conducting project reviews.

MULTIPLE PROJECTS

You may become responsible for the program management of a set of interrelated projects. In that program management role, you would have multiple project managers or task managers reporting to you, so you cannot personally attend all the task reviews and critically examine all the necessary detail. In this case, you must receive some kind of summary information that indicates the status of the several projects (or the many tasks) for which you have responsibility.

Use summary status indicators to get an overall picture, and then probe deeper.

When you cannot personally get into details, you must evaluate the competence and intentions of those who summarize information for you. In the early 1990s, the secretary of defense, a man ultimately responsible for many large developmental projects, had to apologize to Congress (of which he had previously been a member) for misinforming them of the status of a navy stealth attack plane. He had not been fully or accurately informed of the project status (late and overrun) by many of his aides, several of whom then resigned or were reassigned. The project was then canceled. In this case, the advantages of managing multiple projects (you can average your successes and failures, and you may have flexibility to move resources from one project to another) were outweighed by the disadvantages (dependence on summary information, coping with simultaneous problems, and having only limited time to devote to a given project).

Whenever possible, you should personally visit the various project reviews. Imagine that you have three projects reporting to you, each of which is being managed by another project manager, who reviews it monthly. You might sit in on each of these in a rotating fashion, so you attend the review of each project once every three months. If one project you have responsibility for is significantly more important than the others, attend these reviews more frequently (e.g., every other month, if not monthly).

Be visible to the program and project stakeholders.

COMPUTER SOFTWARE

Software is *not* the foundation of project control—it is people and communications. Don't assume that a number is valid just because a computer generated it. Ironically, software adds to the load of details and may increase complexity.

The project manager cannot micromanage and intimidate; he or she needs to establish an approach to focus on what is important and secure the assistance of the project team. Most project participants have a natural desire to avoid thinking about administrative and integrative details and spend time in more comfortable and "fun" areas.

Figure 17-5 is an example of how you might replan the example project shown in Figure 7-18 to make it easier to monitor. In this illustration, two critical path tasks (C and E) that were long (60 and 55 days, respectively) have been subdivided into three shorter subtasks.

TYPICAL PROBLEMS

The word *control* has a pejorative connotation, implying power, domination, or authority. However, by definition, management is a process of directing resources to achieve objectives, so that connotation is not warranted.

(continued)

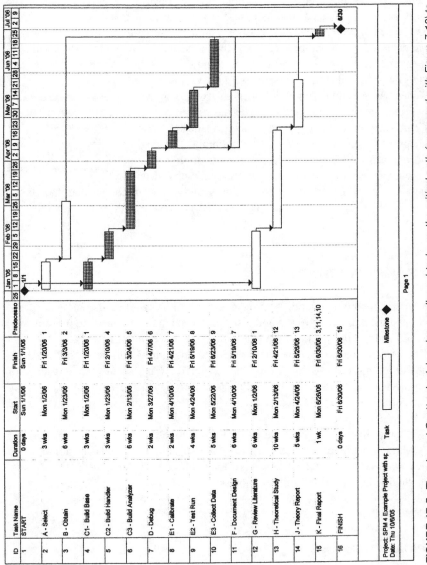

FIGURE 17-5. Time-based Gantt schedule using smaller subtasks on the critical path (compared with Figure 7-18) to improve monitoring.

Problems arise because of too much or too little monitoring. Too much monitoring—in the form of micromanagement—may demotivate personnel, consume too much time, or cost too much. Inadequate monitoring, based perhaps on the naive optimism that the project will be performed in accordance with plan, can lead to disaster. The following (edited) quotation comes from a memo written months after the start of a new product development project:

A. The accessories were far more complex than anticipated at the time the original plan was prepared. There was not, at that time, any defined level of performance for these accessories: the feasibility of the accessories had not been studied. The estimates for the optical systems and mechanical components of the accessories were inadequate until several months of feasibility study had been performed and the products fully defined. Even after they were defined, all of the accessories have had to go through repeated design iterations. In October, all of the accessory designs were rejected. The Head Adapter and the Lamp Housing required mechanical design rework to reduce their size. In addition, the Lamp Housing required optical redesign to accommodate additional optical features necessary for a smaller, more compact package. The Head Adapter was finally approved in mid-December; however, the base was not approved until mid-January. The Lamp Housing received verbal approval only last week. The Accessory, as designed in this project, was found to be completely unacceptable, requiring total redesign, forcing its removal from this project (all future efforts on the Accessory will be covered by a separate project and are excluded from this summary).

The original plan of last February (13 months ago) contained approximately 600 hours of optical and mechanical design for the Accessories. In our revised plan, including labor to date, the optical and mechanical design is approximately 2,400 hours; the difference is 1,800 hours plus drafting time.

B. The original plan called for 670 hours for industrial design. Because the designs have been subjected to repeated revisions, the revised plan, including labor hours spent to date, puts industrial design at approximately 1,700 hours: an increase of 1,030 hours.

C. The layout and design of the Indicator Dial required considerably more time than that provided in the original plan. The Dial design was rejected several times, which added approximately 800 hours.

D. The custom microcircuit for the digital meter version added coordination time with the vendors and with purchasing of approximately 300 hours.

E. It was assumed in the original plan that the proprietary microcircuit would be available as an input to this project. Delays in receiving properly functioning devices made the timing critical. Therefore, coordination with the vendor and evaluation of sample devices of about 400 hours was performed on this project.

F. Due to the complexity of the mechanical design, we added preparation of a tooling plan, preliminary tool design, and coordination to the project. This has added approximately 450 hours. (We will also do the tool design, but that work is covered under a separate project.)

(continued)

G. The revised plan contains 90 hours for planning and supervision; that time is now charged to the project rather than handled as indirect overhead.

H. The original plan was based on burdened labor rates that were in effect when the plan was prepared. The change in burdened labor rates since then has added approximately 20 percent to the project cost.

I. Personnel being shifted temporarily to other projects has caused approximately 200 to 300 hours to be added.

J. Approximately 340 hours were added to support engineering time to better match experience on other projects.

K. The original plan allowed for about 800 hours for project administration. The actual and planned hours now represent an increase of approximately 800 hours due to slightly higher monthly hours plus the longer duration of the project.

L. Compared to the original plan, a higher level of technical documentation hours have been expended. Additional drafting time associated with the Accessory mechanical designs and redesigns (due to size rejections) is approximately 600 hours. Other factors were the low estimates for standard parts drawings and reworking circuit layouts due to mechanical and size limitations. These other factors represent approximately 1,200 hours.

This project has gotten into severe cost (approximately quadrupled) and schedule difficulties, and it probably got that way by slipping one day at a time. While the issue of changing requirements was not explicitly identified, this is probably an important contributing factor. More frequent monitoring could have detected deviations from plan in time to redirect—or terminate—the project. Chapter 20 will examine more on how to manage scope changes driven by changing requirements.

In addition, there is a monitoring problem because cost (and other) report time lags delay news of project difficulties. In some instances, task leaders or other key people may not be motivated to submit progress reports. This problem is not uncommon where these people are technical experts who are more comfortable with their technology than with writing. Further, there are inaccurate reports, even from conscientious people. In other cases, unclear reports will mask the deviation that actually has occurred.

Even if the reports are prompt, accurate, and clear, you may not notice the deviation when you are busy with other urgent activities. Whenever a deviation is noticed, it takes time to react. Finally, you want to avoid "scapegoating," the practice of searching for a party to blame; instead, see it as a "systems" challenge, and work to improve the system.

HIGHLIGHTS

- A simple way to define project control is the acronym BMAC: baseline, measure, assess, and corrective action.

- Comparison with the project plan provides the basis for monitoring a project.
- The first 15 percent of the project is critical, and you need to establish the monitoring early.
- Monitoring by only recognizing tasks as 100 percent complete is the most conservative measure of progress.
- Projects managers may exercise control by requiring their approval or trusting task managers, but the best approach is to examine the status of tasks.
- Understand the communication requirements of your audience.
- Be visible.
- There is a dark side to metrics that includes analysis paralysis and micromanagement.

18

Project Reviews

Reviews are important to project control because they help to ensure that the project team members understand and report status. This chapter describes the general necessity to conduct reviews on a periodic basis as well as a topic-specific basis.

In the preceding chapter, we described control as a process of comparing the status of the actual work to the baseline to evaluate the need for corrective action. We also discussed the differences between status and forecasts. In this chapter, we are going to describe an essential technique: conducting project reviews. Reviews are a method that organizations use to communicate status. This status can foster a better forecast of the end point of the project. Reviews also can surface concerns that will require corrective action.

THE NECESSITY FOR REVIEWS

Every project will depart from plan, and what you don't know in advance is when this will happen.

A review is analogous to an airplane's navigator. The purpose of both is to monitor the status, focusing on unwanted deviations and suggesting corrective actions. Experienced project managers know that the project will not proceed as planned, but they do not know how it will deviate. Only the naive project manager believes that the plan is sufficient and that no further navigation is necessary to arrive at the project's Triple Constraint destination.

Reviews are your off-course alarm.

The project manager's customer, boss, and other senior management will want to know about the project's status. The people working on the project also will wish to have reviews of the overall project from time to time. Reviews with the project team can enhance communication and motivation.

THE CONDUCT OF REVIEWS

Whether reviews are periodic or topical, you should invest some time in determining what information you need, what questions you will ask of whom, and how much time to allocate.

Planning

Project reviews are important tasks that require planning of the scope, schedule, and cost. The goal is acquisition of all relevant information. The schedule may include a simple statement that the review will consume two hours on a particular afternoon. The cost depends on the number of people who participate and the preparation time. Everyone involved should understand the goal and be prepared to carry out their assignments. This means that the project manager must know how much time and what level of detail are appropriate for everyone's participation.

Manage the task of conducting a project review.

Reviews also may be thought of as a particular kind of meeting. As such, all the care of preparation and follow-up discussed in Chapter 16 is relevant.

Questions

Good project managers will ask questions at project reviews in order to discover variances and apply corrective action. You want to learn about what has and has not happened so far so that you can change the future. Whether you like it or not, you cannot change the past. Although questions about the past can help you to anticipate what must be done in the coming days or weeks, they may be intimidating if something has gone wrong. Thus, ask questions nonthreateningly. You are not asking questions to embarrass or pillory anyone. If you (or others such as configuration management or quality assurance) don't ask questions, some people won't volunteer critical information. Examples of good review questions include the following:

You cannot change the past, but you can influence or change what is still to happen.

- What is your greatest concern?
- Do you anticipate any problems that we haven't talked about yet?
- What persistent problems do you have, and what is being done to correct them?
- Do you need any resources (people or things) that you do not yet have?
- Are there any personnel problems now or that you anticipate?
- Do you know of any things that will give you schedule difficulties in completing your task? If so, what are they?
- What kind of help would increase your confidence in the schedule?

- What schedule risks have you considered?
- Is there any possibility that completion of your task will lead to any technological breakthroughs for which patents might be appropriate?

Ask good questions that elicit truthful, constructive responses.

- Has any work done on your task led to any competitive edge that we might use to gain other business elsewhere?
- Is there anything I can do to help?

There are judgmental or backward-looking questions that may be harmful:

- What went wrong?
- Who made that mistake?
- Why didn't you anticipate that?

In most cases, these areas should be explored in private sessions, and even there, the questions can be reworded to be less threatening. For example, the first question might be restated more usefully as, "What would you do differently next time?"

Plan to conduct project review meetings, and expect problems to surface.

The thing to remember about project reviews and these questions is that you are almost assuredly going to hear some bad news. Most of us do not cope with bad news in a very positive way, so the project review easily can become a recrimination and blaming session. This will not be productive. It will destroy the review and much additional effort on the project. Be businesslike and factual in conducting the reviews, and keep asking questions to gather information. Remember, don't kill the bringer of bad news.

PERIODIC REVIEWS

As a general recommendation, every project should be reviewed once a month, if not more often. Periodic reviews can catch deviations from plan before they become major disasters. In the case of the overrun new product development project discussed in Chapter 17, periodic project reviews could have caught the deviations at the end of one or two months, when something might have been done about them, rather than at the end of 13 months, when the accumulated deviation from the plan was so severe.

Task Review

In the most conservative approach to determining project status, a task is in one of two conditions: complete or not complete. For tasks whose performance axis dimension has been completed, examine the actual versus planned cost and schedule, as illustrated in Figure 18-1. Assuming that all tasks were estimated on a uniform basis, for instance, a 50:50 likelihood of underrun or overrun, and unless there is something unique about the cost deviations on any completed activity, the accumulated actual cost versus plan can be used to project the cost at the end of the project. In Figure 18-1, actual cost for the five complete tasks ($42,000) is less than the planned cost ($44,100), and the ratio of these indicates that the final actual costs will be approximately 97.5 percent of the plan.

It is appropriate to identify incomplete tasks that are in progress, as well as those which have not yet been started. The project manager typically determines the project's schedule status first at the critical path and then at the tasks that are not on the critical path. Note that an activity not on the critical path often will slip off the schedule baseline simply because float (or slack) was available that allowed the project manager to let it slip beyond its scheduled start date. For tasks that are under way, find out whether there have already been any difficulties that would preclude their being completed on time. The calculation of "schedule variance" is helpful in making forecasts of the estimated completion date of the project.

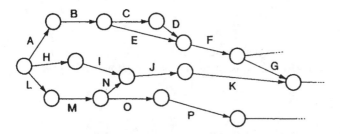

1 - Identify technically complete activities.
2 - Examine actual versus planned cost and schedule:

Completed Activity	Cost			Schedule		
	Actual	Plan	Variance	Actual	Plan	Variance
A	3,000	3,200	200	7	6	<1>
B	4,100	2,900	<1,200>	9	5	<4>
C	6,000	8,400	2,400	6	11	5
D	9,700	12,600	2,900	12	17	5
E	19,200	17,000	<2,200>	24	23	<1>

FIGURE 18-1. *Measuring progress.*

> **Watch the
> schedule, and
> recalculate the
> critical path.**

In the case of the project illustrated in Figure 18-1, the next concern would be the status of tasks H, L, and F. Do these tasks not yet completed indicate that the project is behind schedule? After exploring that, we would want to know about the critical path for this particular project (which is deliberately not shown in the figure). In this case, the concern is whether the completed activities have already caused the project to slip, or whether the schedule variations have no significance with regard to the overall project.

In the case of cost review, it is necessary to examine the details of each task individually, as shown in Figure 18-2. The project manager who looks only at the overall project may be deceived. Project cost can appear to be in harmony with the plan, but that may conceal compensating task overruns and underruns or other difficulties that must be explored. Naive managers assume that actual cost in excess of plan is bad (task A in Figure 18-2), but this may only signify work being completed early. Similarly, actual cost underrunning plan is naively assumed to be good news (task B in Figure 18-2), but this actually may be symptomatic of a situation in which actual progress is inadequate. People may be avoiding the task because it is difficult or unpleasant. To determine both status and forecast, the project manager must ask questions, not just look at project cost data.

Commitments

Project cost reports are always plagued by certain problems. First, they are not "real time"; they are not issued instantly at the end of the month (or any other

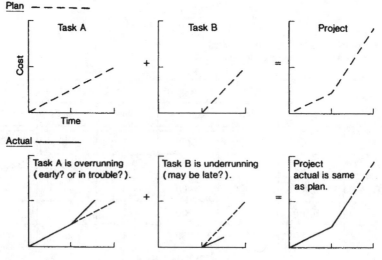

FIGURE 18-2. Details matter.

calendar period). It takes time to gather the information, process it, print it out, and distribute it, whether the system is manual or electronic. Second, these reports are seldom the highest priority in any corporation's accounting department, coming after payroll, reports mandated by law, and usually customer billing. Thus, project cost reports typically follow completion of the monthly period by two or three weeks (at least a week and a half and, in some cases, five or six weeks).

In addition, cost reports are nothing but a record of the apparent charges to a project, which may be in error. Even if the cost charges to the project are correct, they do not necessarily reflect the corresponding amount of perform-ance accomplishment. If the actual costs for each task agree completely with the plan for each task, there is no assurance that the work accom-plishment has a corresponding degree of completion. Within an individual task, where there is no detailed further subtask breakdown, the project manager has only judgment to guide him or her in determining whether progress and cost are con-sistent. However, if a project is divided into many tasks, the ambiguity is reduced because the project manager can look at completed tasks and compare their actual cost with plan. In looking at these costs for tasks, the manager also must look at commitments. Commitments are obligations for which the project will be billed but that are not reflected on the actual cost reports. Figure 18-3 illustrates two common problems.

Commitments will become future costs.

A task apparently completed with an underrun may in fact have outstanding bills charged against it. To preclude surprises, look at the commitments charged against the task. If system-generated reports do not exist, the project manager must main-tain these manually to avoid unpleasant surprises.

To summarize, review the three common dimensions of project performance:

Month	Event	Commitment Report	Cost Report
1	$10K purchase order issued	$10,000	-0-
2	$1K trip authorized	11,000	-0-
3	Traveler paid $900	10,000	$900
4	Vendor delivers	10,000	900
5	Vendor sends bill	10,000	900
6		10,000	900
7	Vendor's bill paid	-0-	10,900

FIGURE 18-3. Commitments are future costs.

- *Technical dimension.* Determine if the work is complete or not.
- *Schedule dimension.* Compare actual and planned time for completed tasks.
- *Cost dimension.* Compare actual and planned cost for completed tasks. Include commitments.

Cost to Complete

Periodically, the project manager should develop forecasts of the cost (and time) to complete the entire project. You might do so quarterly or semiannually, as well as whenever the plan is revised substantially.

Do not make these estimates by simple subtraction of cost to date from the planned cost, as in method 1 in Figure 18-4. Have each task manager reestimate the task in light of everything now known, as shown by method 2 in the figure. Unfortunately, this often indicates that the project will overrun its budget, but it is important to learn of is possibility early enough to do something about it. As the project progresses, the participants learn about what is involved. Also, replanning may occur, and the necessity for new tasks may become apparent.

For example, assume that at the fourth monthly project review you are shown the summary information in Figure 18-5. All tasks that are supposed to be completed actually have been completed, as has task F, and progress is being made on the two other tasks. In addition, actual costs to date for completed tasks are below plan. However, even though there has been no schedule slippage of tasks on the critical path, a careful examination of the schedule shows that the original critical path has been replaced by a new critical path (the rest of task H, plus tasks J and K) owing to the very late completion of task G.

FOLLOW-UP ACTIONS

Follow up reviews.

During any review, a variety of actions will be identified to cope with the various problems uncovered. The project manager always should record these actions, the person responsible, and the expected completion date. This might be done on a register (Figure 18-6). All people concerned and their supervisors should receive copies,

Method 1			Method 2		
Plan (or Allowed) Cost	=	$10,000	Task Q	=	$1,632
Cost to Date	=	6,000	Task R	=	2,149
Cost to Complete	=	$4,000	Task S	=	1,910
			Cost to Complete	=	$5,691

FIGURE 18-4. *Estimated cost to completion.*

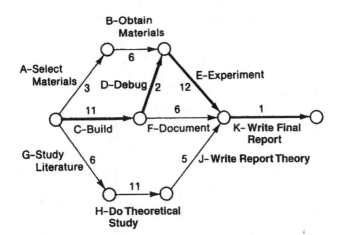

TASK	COST ($)			SCHEDULE (WEEKS)		
	Plan	Actual	Variance	Plan	Actual	Variance
A *	5,250	6,850	<1,600>	3	3	—
B *	9,125	8,725	400	6	8	<2>
▷ C *	87,270	89,890	<2,620>	11	11	—
▷ D *	10,900	10,900	—	2	2	—
▷ E	177,200	P		12	P	
F *	92,400	81,200	11,200	6	5	1
G *	18,220	19,300	<1,080>	5	10	<5>
H	42,020	P		11	P	
J	17,500	N		5	N	
▷ K	14,950	N		1	N	
TOTALS FOR COMPLETED TASKS	223,165	216,865	6,300			

Legend
▷ Critical Path
* Task Complete
P Task in Progress
N Task not Started

FIGURE 18-5. *Summary of a project's cost after four months. Cost projection is for a 3 percent underrun, but the critical path has now changed, and the project will be one week late unless something changes.*

and the status of these action assignments should be reviewed no later than the next project review.

TOPICAL REVIEWS

We are going to use the phrase *topical review* to mean reviews that focus in on some functional or life-cycle activity. The titles for these reviews will vary from

Assigned Actions for Completion

Project: ____________________

Action	Responsibility	Due

FIGURE 18-6. *Typical action follow-up form.*

organization to organization and will differ for various types projects (e.g., software and construction). For example, many new product development, engineering, and other hardware projects have the following types of reviews:

1. Conceptual design
2. Preliminary design
3. Critical design
4. Manufacturing
5. Preshipment
6. Management
7. Customer

Often, contracts specify these reviews, and they may be a precondition to carry out further work on the project. In such a case, the reviews themselves clearly would be designated as task activities on the network diagram and be considered major project milestones, as discussed in Chapter 7.

Management and customer reviews are often onerous, but they may stimulate participation and involvement on the part of all people working on the project. For this to happen, the project manager must solicit ideas from all project personnel as to what should be discussed during the review. Rough ideas for the review should be delineated in a smaller group of key staff. At this point, the manager should delegate portions of the review to other people. (Be sure that there are not so many people making presentations at the formal review that it becomes a circus.) Next, conduct a trial run with a fairly large group of project participants invited to criticize. Following this, prepare materials for the formal presentation, before which it is desirable to conduct a second dry run with a peer management group that represents the same range of skill backgrounds as the audience that will attend the formal review. The people attending this dry-run review will provide additional insights as to how materials can be better presented or changed. If possible, videotape the presentation for later review.

After you conduct the formal review with top management or the customer, it is a good idea to debrief the project team. Team members are just as interested as you are in the kinds of questions that management and the customer asked and will find it just as interesting and motivating to spend some time hearing what happens and getting an overview of the particular issues covered in the review.

Major review presentations can be an opportunity.

TYPICAL PROBLEMS

Reviews are plagued by three common problems. First, there is always a concern as to whether the information being presented or discussed is accurate. Cost reports, as noted previously, are especially prone to error. Beyond this, there is often speculation about the exact status of some task, component, or report. Clearly, good planning for reviews can help you to manage this issue.

The second problem is the poorly conducted review. Aimless discussions or recriminations are common. Running a review like any other well-planned meeting greatly reduces the possibility of getting off the track.

The third problem is that some project personnel fail to meet schedule deadlines and do not consider this to be a major problem. For example, technologists may believe that it is more important to arrive at a perfect or elegant solution (however long it takes) than an adequate solution (on schedule). "Better" is often the enemy of "good enough."

HIGHLIGHTS

- Reviews are a good way to evaluate variation from baseline and other potential problems with expectations of fostering agreement on the needed corrective action.

- Reviews must be planned.
- Ask good questions, and expect problems to surface at project review meetings.
- Periodic reviews should be conducted as appropriate for the project, but once a month is a good rule of thumb.
- The type of review depends on the project and the customer's requirements.
- Questionable accuracy and poor procedures are common problems with reviews.

19

Project Cost Reports

A project cost accounting system is required to monitor the cost baseline. The analysis may help you to identify variances with schedule or performance baselines. We discuss these topics and include examples of cost reports, which normally are generated as computer printout. We also discuss the earned-value management system.

COST MONITORING

The availability of information to monitor a project's actual cost varies among organizations. Any firm that performs projects should have a project cost accounting system. Commercial customers may not choose to audit a cost-reimbursable contract, especially if the value is relatively small, but governmental organizations routinely conduct audits during and after projects. Governmental organizations also may audit project costs on fixed-price projects. Thus, firms doing work for the federal and other governmental entities and many firms in the construction industry have well-developed project cost accounting systems that capture labor and non-labor project costs.

Firms also must treat capital financial investments differently from costs or expenses that can be written off in the current financial period. Therefore, when a firm carries out a project that adds to its capital infrastructure, it must collect and report those costs. A typical situation is the addition of a new production line in a factory, where plant engineering labor costs and other direct expenses are capitalized along with the purchase price of the newly installed production equipment. Such capital costs may be audited by tax authorities. Although the firm's system to collect and record such capital costs may not be as elaborate as the one for contract projects, a system must exist.

Other firms carrying out projects may not have systematic project cost accounting systems. One example is many firms conducting new product development projects. The important issue in such firms is before-the-fact planning of what resources will be required, the time to market, the expected profit to be earned, and the payoff in comparison with alternative investments in other new products. Once the project is under way, costs per se are of relatively little importance compared with staying on schedule. All costs (except any capital expansion to build new or specialized production capacity) are expensed in the current year, and these typically are controlled by limiting the staff headcount level.

Sunk costs do not justify continuing a project.

Many projects, especially new product development projects, are terminated prior to plan or market entry (or later, before the original profit goals are achieved). Sometimes managers argue to continue a project because a large investment already has been made. This is an error because expensed costs are merely written off (although unrecoverable capital costs must be treated differently). Throwing "good money after bad" is never a justification to continue a marginal project.

COMPUTER COST REPORTS

Variance from baseline is a signal to investigate and be prepared to take corrective action.

Many organizations, especially large ones and those performing projects under contract to a governmental entity, commonly have computer-generated reports that summarize project cost. Such reports normally come from a centralized enterprise information system. In some cases, they also may cover schedule deviations. A number of report systems are available for purchase or lease. Figure 19-1 shows how such reports can be useful for project monitoring. Actual cost (and schedule) data are collected from labor time cards, purchase orders, and other direct charges to the project. These are compared with plans, noted variances are analyzed, and required corrective actions then may lead to replanning. Comparing reality with the plan may suggest that certain trends will lead to future variances, which again is a cause for replanning. To be useful, this comparison must be done for each cost center (e.g., department) and work breakdown task.

The project manager must steer a careful line between having too many forms and too much information and having too little information to monitor the project, although it is probably better to have more information than less. Neither the project manager nor top management should be trapped into believing that actual cost data (which can be reported with great precision) are the only measure of project health (which requires difficult three-dimensional measurements).

Most organizations with computer support for project management issue a weekly labor distribution report. This report lists the names of people charging time to each project during the previous week. This key report for project managers provides an early warning signal. Examination of the report may reveal that people are charging your project who should not be charging it and that people whom you expected to be working on it are not charging your project.

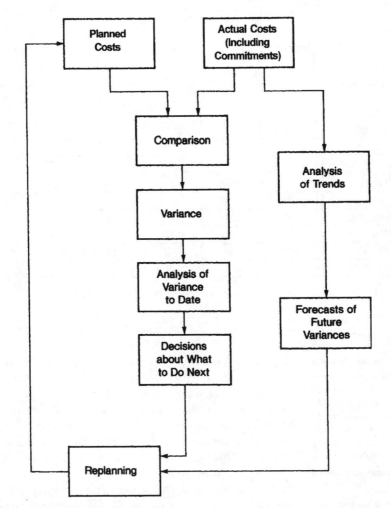

FIGURE 19-1. *Replanning is called for whenever future trends or past actuals indicate significant deviation from plan.*

COST MONITORING PROBLEMS

This section reviews a few of the cost reports for an example "materials study project." Actual report formats vary in different organizations, and the level of detail can range from minimal to extensive. (Figure 19-2 is one example of the kind of information that would be provided by a reasonably helpful cost reporting system.) These reports often are prepared by a centralized project cost accounting function. It is not always necessary to use a computer on a small project, although the amount of data to be handled otherwise can be a substantial burden, even to a well-qualified project cost accountant.

Project Cost Report

Project *Example*   **Task** _B_   **Department** _456_

Category		Month	1-Jan		Totals To Date		
		Plan	Actual	Variance	Plan	Actual	Variance
Hours	Sr. Prof.	40	40	0	40	40	0
	Jr. Prof.						
	Sr. Tech.						
	Jr. Tech						
Dollars	Labor	1,000	1,000	0	1,000	1,000	0
	Overhead	1,000	1,000	0	1,000	1,000	0
	Nonlabor	4,400	0	4,400	4,400	0	4,400
	Prime Costs	6,400	2,000	4,400	6,400	2,000	4,400
	G & A	960	300	660	960	300	660
	Total Costs	7,360	2,300	5,060	7,360	2,300	5,060

FIGURE 19-2. Cost report for first month for task B work by Department 456.

Variances Due to Timing

A variance between actual cost and plan may be due to payments being made later than plan rather than to work variances.

Figure 19-2 is the project cost report for task B within a business unit (Department 456) at the end of the first month. Typically, this report would be available about the middle of the second month (February, in this case). In this case, labor and the overhead thereon are in accordance with plan. However, there is a favorable variance in the nonlabor expense. That is, there was a plan to spend $4,400, but nothing has been spent.

Figure 19-3 shows the corresponding task and period commitment report for the same business unit. Commitments in the amount of $4,400 have been incurred.

Project Commitment Report

Project *Example*   **Task**   _B_   **Department** _456_

Month   _1_

Commitment Date	Item	Amount	Estimated Payment Date
21 Jan	P.O. - Materials Supply Co.	$ 4,000	30 Apr
21 Jan	Travel Auth - C. Williams	$ 400	15 Feb
		$ 4,400	

FIGURE 19-3. Commitment report for first month for task B work by Department 456.

Thus, the cost variance merely indicates that bills have not yet been paid rather than being a variance because of activities not yet undertaken. To put it another way, it is anticipated that the $4,400 expense will occur later. This demonstrates that it is impossible to make intelligent use of project cost reports without also examining commitment reports.

There also may be variances caused by purchasing a large amount of some material, some of which is needed for an early task and the rest for a later task (e.g., to obtain a volume discount). In this situation, the receipt (and cost) of material does not conform to task usage.

Figures 19-4 and 19-5 are the same reports for the end of the second month. Once again, the labor hours and costs, as well as the overhead, are in accordance with the plan. In this case, there is an unfavorable variance during the second month with regard to nonlabor costs because the travel voucher is paid now, but the plan had the expense in the first month. For the totals to date, that is, through the end of the second month, the nonlabor variance is favorable. This favorable variance is composed of the unpaid $4,000 purchase order, which is a variance only because of payment timing, and a $50 favorable variance because the travel voucher payment was $50 less than plan.

Variances Due to Actual Work Not as per Plan

Figures 19-6 and 19-7 are the project cost reports at the end of the third and fourth months for the same task. In this case, labor hours exceed plan in the third month, with attendant unfavorable cost variances, which happen to offset exactly the pre-

Project Cost Report

Project *Example*   **Task** *B*   **Department** *456*

Category		Month 2-Feb			Totals To Date		
		Plan	Actual	Variance	Plan	Actual	Variance
Hours	Sr. Prof.	4	4	0	44	44	0
	Jr. Prof.						
	Sr. Tech.						
	Jr. Tech						
Dollars	Labor	100	100	0	1,100	1,100	0
	Overhead	100	100	0	1,100	1,100	0
	Nonlabor	0	350	(350)	4,400	350	4,050
	Prime Costs	200	550	(350)	6,600	2,550	4,050
	G & A	30	83	(53)	990	383	607
	Total Costs	230	633	(403)	7,590	2,933	4,657

FIGURE 19-4. *Cost report for second month for task B work by Department 456. (Note: Numbers in brackets are unfavorable.)*

Project Commitment Report

Project *Example*   Task *B*   Department *456*

Month   *2*

Commitment Date	Item	Amount	Estimated Payment Date
21 Jan	P.O. - Materials Supply Co.	$ 4,000	30 Apr
21 Jan	Travel Auth - C. Williams	Paid	Paid
		$ 4,000	

FIGURE 19-5. *Commitment report for second month for task B work by Department 456.*

vious favorable variance on nonlabor because of the travel variance. In the fourth month, the purchase order is paid, and the net variance for the task becomes zero.

Actual work may differ from plan.

Figures 19-2 through 19-7 illustrate that variances occur because of the payment timing and that actual performance differs from plan. They also indicate the necessity of examining the details in project cost reports and commitment reports to understand the reported variances and their significance.

Each accounting system differs in detail, so the project manager should understand exactly how the reports are prepared (i.e., to what errors the reports are prone) as well as the specific meaning of each column of data.

Project Cost Report

Project *Example*   Task *B*   Department *456*

Category		Month	3-Mar		Totals To Date		
		Plan	Actual	Variance	Plan	Actual	Variance
Hours	Sr. Prof.	2	3	(1)	46	47	(1)
	Jr. Prof.						
	Sr. Tech.						
	Jr. Tech						
Dollars	Labor	100	125	(25)	1,150	1,175	(25)
	Overhead	100	125	(25)	1,150	1,175	(25)
	Nonlabor	0	0	0	4,400	350	4,050
	Prime Costs	200	250	(50)	6,700	2,700	4,000
	G & A	30	38	(8)	1005	405	600
	Total Costs	230	288	(58)	7,705	3,105	4,600

FIGURE 19-6. *Cost report for third month for task B work by Department 456.*

Project Cost Report

Project _Example_   **Task** _B_   **Department** _456_

	Category	Month	4-Apr		Totals To Date		
		Plan	Actual	Variance	Plan	Actual	Variance
Hours	Sr. Prof.	0	0	0	46	47	(1)
	Jr. Prof.						
	Sr. Tech.						
	Jr. Tech						
Dollars	Labor	0	0	0	1,150	1,175	(25)
	Overhead	0	0	0	1,150	1,175	(25)
	Nonlabor	0	4,000	(4,000)	4,400	4,350	(50)
	Prime Costs	0	4,000	(4,000)	6,700	6,700	0
	G & A	0	600	(600)	1,005	1,005	0
	Total Costs	0	4,600	(4,600)	7,705	7,705	0

FIGURE 19-7. Cost report for fourth month for task B work by Department 456.

Variances Due to Overhead Rate Changes

Figure 19-8 is the project report for task E in a different business unit (Department 789) as reported at the end of the fourth month. In this report, there are favorable variances in labor but an unfavorable variance in overhead. How can this be? If labor is favorable, why should overhead be unfavorable?

Project Cost Report

Project _Example_   **Task** _E_   **Department** _789_

	Category	Month	4-Apr		Totals To Date		
		Plan	Actual	Variance	Plan	Actual	Variance
Hours	Sr. Prof.						
	Jr. Prof.						
	Sr. Tech.	200	120	80	200	120	80
	Jr. Tech	400	400	0	400	400	0
Dollars	Labor	7,000	5,800	1200	7,000	5,800	1200
	Overhead	7,000	7,540	(540)	7,000	7,540	(540)
	Nonlabor	0	0	0	0	0	0
	Prime Costs	14,000	13,340	660	14,000	13,340	660
	G & A	2,100	2,001	99	2,100	2,001	99
	Total Costs	16,100	15,341	759	16,100	15,341	759

FIGURE 19-8. Cost report for fourth month for task B work by Department 789.

**Factors outside the
manager's control
may cause costs
variances.**

Figure 19-9 shows the cause of this. A planned overhead rate (namely, 100 percent) was based on a planned workload for the technical support section (or perhaps the entire research and engineering division). However, the actual overhead rate at the end of the fourth month is higher (namely, 130 percent) because the workload base for the entire section has been reduced from plan. The overhead expenses have been somewhat reduced, but not in the same proportion as the direct labor because overhead is partially composed of fixed expenses that cannot be reduced. Thus, the actual overhead rate turns out in this case to be 130 percent rather than the planned 100 percent. (Such a change in overhead is extreme; we use it simply to dramatize the possible effect of overhead rate changes.) The variances could be summarized as follows:

A senior technician was planned full time but was not released from his previous project at the start of month 4. If labor is not added, the project will be late.

Overhead is now 130 percent, not 100 percent, as planned. This will cause a cost overrun unless compensating savings can be found.

Figure 19-10 is the project cost report for this task in the technical support department at the end of the fifth month. The senior technician category continues to have a favorable variance, which is partially offset by an unfavorable variance in the junior technician category and the continuing unfavorable variance in overhead. The overhead variance again is attributable solely to the overhead rate now being 130 percent rather than 100 percent, as per plan. Examination of these variables might lead to the following kind of information:

Planned

Project Example $100,000 Labor Cost	Project Otherone $300,000 Labor Cost
Overhead ($200,000 Fixed + $200,000 Variable) $400,000	= 100%

Revised

Project Example $100,000 Labor Cost	Project Otherone $150,000 Labor Cost
Overhead ($200,000 Fixed + $125,000 Variable) $325,000	= 130%

FIGURE 19-9. *Cause of unfavorable overhead variance.*

Project Cost Report

Project *Example*   **Task** *E*   **Department** *789*

Category		Month	5-May		Totals To Date		
		Plan	Actual	Variance	Plan	Actual	Variance
Hours	Sr. Prof.						
	Jr. Prof.						
	Sr. Tech.	40	20	20	240	140	100
	Jr. Tech	80	120	(40)	480	520	(40)
Dollars	Labor	1,400	1,500	(100)	8,400	7,300	1,100
	Overhead	1,400	1,950	(550)	8,400	9,490	(1,010)
	Nonlabor	0	0	0	0	0	0
	Prime Costs	2,800	3,450	(650)	16,800	16,790	10
	G & A	420	518	(98)	2,520	2,519	1
	Total Costs	3,220	3,968	(784)	19,320	19,309	11

FIGURE 19-10. *Cost report for fifth month for task B work by Department 789.*

The senior technician is sick, and this category of labor is below plan.

A junior technician, previously unplanned, has been added and has been able to complete the technician work on schedule.

Overhead is still 130 percent, but final dollars are okay.

The junior technician was able to accomplish in fewer hours the work previously planned for a senior technician (which indeed can happen). This net effective saving in labor hours was sufficient to compensate for the increased overhead dollars. The end result in this case is a small, favorable variance.

EARNED-VALUE MANAGEMENT

The earned-value management approach is widely known in the project management profession but is practiced mostly in the government contracting area or on large construction projects. Proponents of the method consider it the best way to achieve an integrated comparison of work performance, schedule performance, and cost performance. However, many individuals and firms have failed to adopt the method for several reasons:

- It requires the company to have a system for project accounting.
- The accounting system needs to able to provide accurate, consistent, and timely reports of expenses.

- It requires some sophistication in measuring partially completed tasks, and this is particularly challenging on projects that involve intangibles such as services or software.

Earned value is a powerful technique

- It requires self-discipline on the part of the project manager, the project team, and the project sponsor.

Three important measurements are used in earned value, and we show them in Figure 19-11. The *budgeted cost of work scheduled* (BCWS) is the baseline, and it is compared with the *budgeted cost of work performed* (BCWP) for a given period of time. The difference between these is termed the *schedule variance*. The *actual cost of work performed* (ACWP) is compared with the budgeted cost of the work performed, and the difference, if any, is termed the *cost variance.*

Here is an example of how earned value would work. Imagine that you have a three-month task and that you expect it to require a uniform rate of effort with costs equal to $3,000 per month. Also, imagine that the task could be divided into three one-month subtasks, as shown in Figure 19-12. Figure 19-13 portrays three options you might use to describe your plan for use with an earned-value system. Table 19-1 summarizes the three quantities and the two variances for each option at the end of each month. In option A, a so-called back-loaded project plan, you have unfavorable variances until the end. In option C, a so-called front-loaded project plan, the situation is favorable until the end. This illustrates that the kinds of estimates used in constructing the plan can conceal schedule and cost variances.

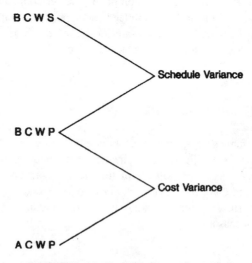

FIGURE 19-11. *Earned-value methodology.*

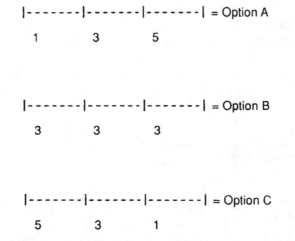

| - - - - - - - | = One-month duration

| - - - - - - - | - - - - - - - | - - - - - - - | = Total task
   3   3   3

FIGURE 19-12. *A three-month task with planned costs of $3,000 per month.*

| - - - - - - - | - - - - - - - | - - - - - - - | = Option A
   1   3   5

| - - - - - - - | - - - - - - - | - - - - - - - | = Option B
   3   3   3

| - - - - - - - | - - - - - - - | - - - - - - - | = Option C
   5   3   1

FIGURE 19-13. *Three options for describing your plan for use with an earned-value measurement system.*

TABLE 19-1 Analysis of Earned-Value Exercise Options

Option	BCWS	BCWP	ACWP	Schedule	Budget
A					
End of Month 1	1	1	3	0	−2
End of Month 2	4	4	6	0	−2
End of Month 3	9	9	9	0	0
B					
End of Month 1	3	3	3	0	0
End of Month 2	6	6	6	0	0
End of Month 3	9	9	9	0	0
C					
End of Month 1	5	5	3	0	+2
End of Month 2	8	8	6	0	+2
End of Month 3	9	9	9	0	0

COMPUTER SOFTWARE

The possibility of comparing actual versus planned costs is a frequently advertised feature of computer software. As we said earlier, it is relatively easy to insert expected costs (labor and nonlabor) for each task when it is entered into a computer project management software package. It is also possible during project performance to enter actual costs information. However, most large corporations collect these actual cost data in the corporate computer (or a dedicated project management computer). That is, each cost center (e.g., different departments) working on the project provides data that are transmitted from their location (often remote from the project manager). Although it is possible to get a report from the corporate computer and enter these data in your desktop computer, it is simpler to have an electronic data link, as illustrated in Figure 19-14. In fact, several project management computer software packages will trade data back and forth with certain project management systems.

Most high-end project management computer software packages are capable of linking to a corporate cost collection system. Alternatively, a project manager might use a simple spreadsheet.

TYPICAL PROBLEMS

Cost reports are late and contain some degree of inaccuracy.

In addition to management's tendency to presume that cost status is the only measure of a project's progress, a problem mentioned previously, there are two other common problems. First, cost reports are never available immediately after the end of the reporting period (typically the fiscal or calendar month). They usually are issued about two or three weeks after the close of the report period. Thus, there is always a time lag with which the project manager must contend. This problem cannot be solved but must be accepted because of other accounting priorities. An alert project manager can recognize the situation and make full use of other available data to stay more current, including weekly labor distribution reports (indicating who did or did not charge time to the project), purchase orders, travel vouchers, and drawing releases.

The second problem is that cost reports often have errors, such as charges that should have been allocated to other projects or overhead accounts. The project manager or an administrative assistant must study the reports carefully and not merely accept them as gospel.

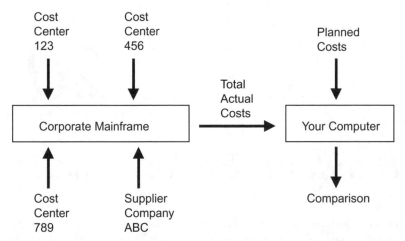

FIGURE 19-14. *It is easier to compare actual versus planned costs if your corporate project cost accounting system can provide actual cost information via a data link to your project management computer software. In some cases, this linkage may include costs of supplier companies.*

HIGHLIGHTS

- A project accounting systems is useful. It should quickly and accurately report project data.
- Earned value is a powerful technique for the integrated analysis of project status.
- Computer-generated cost reports show variances from plan, which usually require corrective action.
- Reports can show variances owing to timing, actual work deviating from plan, or overhead rate changes.

20

Handling Project Changes

Because of the dynamic nature of projects, project managers need to recognize potential changes to the project's baseline and manage them. This chapter provides a systematic approach to managing change.

In Chapter 2, we said that an important project management responsibility is to manage expectations. In Chapters 5 and 17, we discussed the development of a baseline and the project manager's responsibility to measure performance and take appropriate corrective actions. In Chapter 11, we refined the discussion of project planning by describing a systematic approach to managing risks and issues. Regardless of our conscientious and professional performance with regard to project control, requirements will change. We need to remember that all projects involve agreements, and we need to adjust our agreements. As an example, let's consider this situation:

> A married couple—John and Juanita—decide to build a new home for themselves and their one child. On 15 January, they sign a contract with a builder and reach agreement to build a two-bedroom home that will meet their current needs for a price of $100,000 with a delivery date of 15 April. On 15 February, Juanita announces that she is pregnant with triplets. This causes them to reconsider their requirements, and they decide to enlarge the house from two bedrooms to four bedrooms. Because the contractor is not going to provide the extra free work, they renegotiate the contract and agree to pay a total of $170,000 and extend the delivery date to 15 May. On 15 March, John hurts his back, and his doctor recommends whirlpool treatments, so the couple decides to add a whirlpool to their bathroom. Consequently, John and Juanita agree with the contract to pay a new price of $180,000 for constructing their home and agree to take possession on 15 June.

Was this project over budget? From the customer's perspective, the answer is "maybe." However, since the requirements changed measurably, the concept of a project budget becomes irrelevant. For the contractor, we can't tell whether or not there was a cost variance, because we don't have the cost baseline data. In other words, we can't tell whether or not the contractor made his or her target profit margin.

Was this project delayed? Did it suffer from scope creep? The answer to each of these questions is "no," because the changes were agreed to.

A PROJECT PERFORMANCE TRACK RECORD: GOOD OR BAD?

One of the driving factors for the increased attention and professionalization of project management is data comparing the initial estimate with the final cost or time. For example, the widely quoted Standish Group study of an information technology (IT) project reported that less than 10 percent of IT projects are completed on time, in budget, and deliver to the customer that which the customer values. Table 20-1 provides some similar data. Although not illustrated in Table 20-1, the data presented by Mansfield and colleagues also show that the situation is somewhat worse than average for large projects and better than average for small projects. This seems to make intuitive sense, because it is more difficult to accurately estimate ambitious undertakings.

Estimates for smaller projects are more accurate.

TABLE 20-1 Time and Cost Overrun Data, Expressed As Multiple (X) of Plan

Project Type	Time	Cost	Source
50 new products (new chemical entities, compounded products, or alternative dosage forms) in ethical drug firm	1.78X	1.61X	E. Mansfield et al., *Research and Innovation in the Modern Corporation,* W. W. Norton, New York, 1971, p. 89
69 new products in proprietary drug laboratory	2.95X	2.11X	E. Mansfield et al., op. cit., pp. 102 & 104
20 management information systems projects	2.10X	1.95X	R. F. Powers & G. W. Dickson, "MisProject Management? Myth, Opinions, and Reality," *California Management Review,* XV, no. 3, 147–156 (Spring 1973)
34 Department of Defense systems from "planning estimate" from "development estimate"	— —	2.11X 1.41X	G. R. McNichols, D.O.D. Report (November 1974), as quoted in R. A. Brown, "Probabilistic Models of Project Management with Design Implications," *IEEE Trans. Engr. Mgmt.,* vol. EM-25, no. 2, 43–49 (May 1978)
10 major construction projects completed 1956–1977	—	3.93X	W. J. Mead et al. (1977), as quoted in E. W. Merrow et al., "A Review of Cost Estimation in New Technologies," Rand Corporation Report R-2481-DOE (July 1979), p. 38
10 energy process plants	—	2.53X	E. W. Merrow et al., op. cit., p. 87

This is another reason to break a large project down into many small tasks. It will be easier to estimate a small task accurately. You also have nature on your side in the form of the central limit theorem. This theorem predicts that the percentage error will be smaller if a larger number of estimates, each with small random errors above or below the estimate, are made.

Change is a constant on projects.

However, it is a mistake to declare that a project initially estimated at X is performed poorly if the final price or schedule is $2X$. Just like the couple in the preceding example, projects have all kinds of events that are difficult to anticipate and require flexibility in order to satisfy the customer's requirements. Further, we need to note that the market, regulatory, and competitive environment has become increasingly turbulent in the past decade; thus, it is not realistic to assume that most long-term projects will have stable requirements.

It is worth noting, too, that customers frequently make decisions to scale back their original vision for an ambitious project as they get a better idea of their requirements and of the time and funding limitations.

THE PROCESS OF MANAGING CHANGES

The home-building example described earlier in this chapter involved the common occurrence of changing requirements. Here are some other examples: a request for increased accuracy in a sensor, a request for extra flexibility in a computer program, a request to add traffic islands on a bridge, or perhaps a request for an additional flight experiment on a space payload. New environmental, health, and safety regulations may introduce new requirements, too.

Good project managers address changes constructively; they should be recognized, defined, evaluated, and controlled. The control approach of "big MAC" (baseline, measure, assess, corrective action) that we described in Chapter 17 applies. Thus, we have a basic, straightforward process for change management:

- During scope management and risk management planning activities, anticipate the areas of likely change.
- During execution, recognize those things that are outside the scope and risk management baseline.
- Evaluate the impact of the change on the project's performance, and determine if the impact is likely to be inside or outside of risk tolerances and contract boundaries. If the answer is yes, continue with the change management process, and if the answer is no, add the change to the issues management list—but do not start work on the change until the baseline is properly modified.
- Develop a change proposal explaining the condition and the justification for changing the contract.
- Negotiate the change.
- Adjust the baseline as agreed.

Figure 20-1 is another variation of a task authorization form, which allows for changes. As we said in Chapter 6, each organization has its own detailed version, but the essential elements are a description of the work and a place for the person authorizing the revised work and the person accepting responsibility for the revised work to sign. This form thus constitutes a "contract," defining in writing

TASK AUTHORIZATION			PAGE OF		
TITLE					
PROJECT NO.	**TASK NO.**		**REVISION NO.**	**DATE ISSUED**	
STATEMENT OF WORK:					
APPLICABLE DOCUMENTS:					
SCHEDULE **START DATE:** **COMPLETION DATE:**					
COST:					
ORIGINAL	ORIGINATED BY:	DATE:	ACCEPTED BY:		DATE:
	APPROVED BY:	DATE:	APPROVED BY:		DATE:
	APPROVED BY:	DATE:	APPROVED BY:		DATE:
REVISION 1	ORIGINATED BY:	DATE:	ACCEPTED BY:		DATE:
	APPROVED BY:	DATE:	APPROVED BY:		DATE:
	APPROVED BY:	DATE:	APPROVED BY:		DATE:
REVISION 2	ORIGINATED BY:	DATE:	ACCEPTED BY:		DATE:
	APPROVED BY:	DATE:	APPROVED BY:		DATE:
	APPROVED BY:	DATE:	APPROVED BY:		DATE:

FIGURE 20-1. *A task authorization form to authorize task revisions.*

the agreement reached to authorize a revised task. In the case of a subcontractor, a revised or changed subcontract document authorizes the revised task.

Communicate the changes to others, as appropriate.

This process may seem to be a lot of work, so people might be reluctant to make a formal plan change. Such a change not only requires work to issue plan revisions, but it also forces us to admit that we were wrong (in the original plan), and often this brings unwanted attention from higher management. Figure 20-1 shows a one-page form, but an actual task authorization, in common with a subcontract, might be many pages long. You should distribute copies to the initiator, the task manager, and the project cost accounting section. A large project may generate many of these, and the amount of time it takes to issue them may be so great as to advise that there first be telephonic or other speedy notice of forthcoming changes. Hence, the task authorization documents used initially to authorize work are also a major change control document.

In the long run, successful project managers find it far less onerous to take the time to make sure that each agreement with people working on the project has been changed than to discover later that some people have been working according to their prior understanding of the project plan. The sponsor and other upper management often can be a valuable ally in providing resources and issues resolution.

UNMANAGED RISKS AND ISSUES

The approach to managing change depends on the type of contract and the project's risk management approach. In a lump-sum contract, the delivery system owns most of the risks and issues and reflects the management of risks and issues in its contract price. In time-and-materials contracts, the ownership of risks and issues depends on the discussions leading up to the agreement. You may want to review Chapter 4 on contract types and Chapter 11 on managing risks and issues. Here are some examples of risks and issues that would stimulate change:

- The customer may be late getting you essential information, as one project manager complained: "In order to complete a major project, I need important and timely data from my customer. Even when the customers do not provide the data, they still demand that I complete the project in a . . . timely manner."
- If the customer's plant is shut down by a strike or for some other reason, and the customer does not wish to receive the project output on a stipulated date, there is typically a cost consequence.
- The customer or the sponsor may impose a change in project funding. Although a delay in funding to the project organization may not appear to change the total available funding, it typically is accompanied by a schedule rearrangement, which normally leads to undesirable cost consequences.
- Inflation may be different from that forecast in the estimating assumptions.

- There are changes in resource availability, either people or facilities. Perhaps the person assigned to a key task is not performing as expected, which is a common problem illustrated by another project manager's comment: "The assigned person charged over two weeks' time to my project and never even installed the compiler on her machine. I was constantly asking her about her progress, but she always managed to make it sound like things were moving along."

A high-quality estimate will include an analysis of risk, with some providing a response plan for it. The rush to prepare a proposal and submit it in accordance with the bidding requirements may preclude there being sufficient time to do a good job of estimating. As another example, your company's management may insist on a plan for an internally sponsored project so that you do not have sufficient time to do a thoughtful job. There is so much inherent uncertainty in some tasks of some projects that a poor estimate is almost a foregone conclusion.

Make sure that your contracts and risk-response plans recognize that projects conditions are dynamic; there will be change.

Many jobs are proposed with deliberate underestimates of the amount of time or money it will take to perform them. This is the so-called buy-in or low-balling situation that we described in Chapter 3 in which a bidder attempts to win a job by making a low bid. It may result from a deliberate attempt to make optimistic assumptions about all the uncertainties in the proposed project, as well as to omit all contingency from the estimates. In a sense, the bidder is making a best-case estimate of time and cost, not realizing that their confidence is a very low percentage. It is better to make realistic estimates.

Uncertainty and buy-ins can cause poor estimates.

Buy-in bids are much more prevalent where the contemplated contract will be in a cost-reimbursable form and the bidding contractor will not bear the financial burden of having made a low bid. They also can occur in a fixed-price contract situation where the bidding contractor is confident that the customer will request changes in scope. Such changes will provide a "get well" opportunity: Increases in both time and cost for the main project can be added onto or concealed in renegotiations necessitated by changes of scope requested by the customer.

TYPICAL PROBLEMS

If you change project (or task) managers during a project, you usually will be confronted with other changes, typically unfavorable. The new person may be unfamiliar with the user environment and with team members' capabilities. The

(continued)

overriding issue, however, is that the new project or task manager did not plan the work for which he or she is now responsible—this is a violation of the golden rule. If you must change the project or task manager during the course of the activity, allow that person some time to review the prior plan and try to determine how to carry out the remainder of the work.

There is always a reluctance to tell the customer and your boss that a revolting development (such as the discovery of unexpected noise in an amplifier or an obscure "bug" in software) has occurred and many reasons to justify delay. Nevertheless, you should deliver the bad news carefully, thoughtfully, and promptly, before someone else does.

A second problem is that task authorizations often are verbal rather than written. Because they promote misinterpretation, verbal authorizations should be avoided. Nevertheless, they are employed in the real world of project management. When you must use them, be sure that you are clear, ask for feedback, and then try for written confirmation.

The third problem with changes is their impact on resource allocation. There is nothing to do but face up to the reality that resources must be rescheduled, as inconvenient as this may be.

HIGHLIGHTS

- Projects are dynamic and require attention to recognize potential changes.
- Projects involve agreements that get formalized into some contract type that encompasses risk.
- You need to recognize and manage work that responds to changed requirements.
- Authorization documents can be used to communicate planning and change control.
- Three problems changes can cause are that managers may be reluctant to inform the customer and higher management of them, verbal authorizations often cause misunderstanding, and resources must be reallocated.

21

Solving the Inevitable Problems

Successful projects have the ability to anticipate, recognize, and resolve problems in an effective and efficient way. This chapter will help you handle problems that your project encounters during execution. Good decisions come from good information. First, we discuss some approaches to coping with problems. Then we describe decision trees, a powerful analytic technique for structuring decisions in light of risk. Following that, we review use of matrix arrays and then discuss problem-solving styles.

THE GENERAL APPROACH

The essence of project management is making decisions. Because of the complexity of and opinions about project "facts," the project manager often finds himself or herself working to bring people to a common understanding of the issues. Problems are inevitable, and the faster you can find and solve problems, the faster you will achieve success. In the following paragraphs, we describe a structured approach to problem solving.

Approach problem solving systematically.

State the Real Problem

One key to problem solving is to identify the root cause, that is, to understand the real problem, and to work toward remediating the root cause rather than spend time and energy on treating the symptoms. Smoke may be emerging from the hardware you built, or the computer may refuse to obey a subroutine command, but the actual problem may be an overheated

A problem well defined is a problem half solved.

component or an improper line of code in the computer program. You will have to decide how and perhaps why these particular problems occurred.

Throughout this book, we have emphasized the importance of requirements as a basis of project success. A requirement is a statement of the problem that is free of the design. In other words, it describes "what" and not "how." If problems arise, revisit your requirements specifications.

Gather the Relevant Facts

A fact-gathering activity helps to clarify the problems. You need to roll up your sleeves and talk to people, dig into published data, and work at separating fact from speculation.

When we expand the boundaries of analysis, we ensure that we are solving the right problem. Good analysts first practice divergent thinking to expand their understanding of the nature of the situation. By expanding the boundaries of analysis, they improve the chances of correctly defining needs.

Next, the good analyst practices convergent thinking to choose the best description of the need. Be on guard for the cultural bias of moving quickly to design, skipping the creativity of divergent thinking.

People trained in engineering and sciences sometimes find themselves in "analysis paralysis." Although it may take a good deal of time to locate information sources, there is also a law of diminishing returns. Because you will never have 100 percent certainty of obtaining all the information, you must learn to exercise judgment as to when to truncate a search for additional information. At that point, you begin to converge on a solution using the information already gathered.

Propose a "Straw Man" Solution

The concept of a "straw man" comes from agrarian societies where a scarecrow— a figure of a person—was constructed of straw and scraps of clothing in order discourage birds and animals from eating crops. A strong breeze could easily destroy the scarecrow, so a straw man is something that is visible and provokes a reaction but can be destroyed by a strong argument or idea.

Often, as you work toward discovery of the root cause, one or more possible solutions quickly come to mind. Some people adopt the first solution that comes to mind, possibly leaping out of the frying pan and into the fire. Admittedly, the pressures to come up with a solution quickly are great. No one likes to walk into his or her boss's or customer's office and say, "We have a problem." Such a crisis generates psychological pressure in the project manager to come up with a solution quickly so that he or she can say, "We have a problem, but don't worry about it because we have a pretty good solution in mind."

A pitfall to watch for is "satisficing," the common shortsighted shortcut of "picking the first acceptable solution rather than seeking out alternatives and striving for an optimal solution." (Economist Herbert Simon coined the word by merging the words *satisfy* and *suffice*.) Projects often need to rework or continuously improved on these "satisficed" solutions.

Emile Auguste Chartier stated, "There is nothing more dangerous than an idea when it is the only one we have." Sometimes the first solution turns out to be the best solution, but experience shows that there is usually an even better solution to be found after just a little more analysis. However, it is better to generate alternatives, as we will describe in the next section.

Develop Several Alternative Solutions

You can improve your problem-solving skills by developing several alternative solutions and then evaluating them. Thus, when a problem has arisen and must be reported, the successful project manager will say, "We have a problem. We may have a possible solution, but I am going to take three or four days (or hours or weeks, as the case may be) to consider other alternatives. Then I will report to you on the options and our recommended course of corrective action." Although such an approach to reporting the bad news may make you uncomfortable initially, it invariably is associated with reaching better solutions.

Developing alternatives enhances the chance of developing an optimal solution.

Select the Best Alternative

After deciding what the best alternative is, you must adopt a course of action.

Tell Everyone—Communicate Appropriately

As an effective project manager, you have earlier made certain that everybody involved in your project knew the original project plan. Now, because you have made changes, you must tell everyone what the new plan is. If you fail to do this, there will be some people working in accordance with an old and irrelevant project plan that will cause additional rework and stress.

Audit the Outcome—Learn from the Outcome

As you implement the best alternative solution, watch how it is working out. As you learn more about the problem you are solving and the approach you have adopted, a better alternative may become clear, which may necessitate a further change in the plan. Auditing will improve your ability to solve problems by showing you how your solutions actually work out.

DECISION TREES

Choosing the best alternative often requires estimating the possible outcomes and their probabilities. An organized way to cope with the situation is to use a decision tree. Chapter 11 provided an extensive discussion of the identification and quantification of risk. The decision tree tool is simple, powerful, and widely used.

Decision trees require you to quantify the outcome.

Here is an example: Consider the decision whether to go to a movie or walk on the beach. This decision and its possible outcomes are shown in Figure 21-1.

Each possibility has chance future events, which have to do with the quality of the movie or the weather. There are also different outcomes, which are illustrated. Outcomes such as those shown, which have more than one dimension to them (happiness and money in the figure), must be reduced to a single numeric value. This can be done by utility or preference techniques.

Figure 21-2 is a general representation of the kinds of decision trees with which you must work. A decision tree always starts with a decision for which there are two or more possible choices. Each choice may be followed by chance future events or subordinate choices (decisions) in any order and with two or more branches following each node. There is a single numeric value outcome, which is typically the present value of the cash flows along the various branches to that outcome.

Decision trees are used routinely in many situations, such as the decision whether to drill an oil well at a given location or to locate a new warehouse in a new geographic region. In both these typical situations, the correctness of the decision depends on the probabilities of things happening in the future (oil being found or business growth), and decision trees are designed to maximize the likelihood of choosing the correct course of action.

Consider the following situation, which often confronts a project manager: You have just received an unsolicited request for a proposal that will yield a $300,000 before-tax profit if you can win the job. Checking with the marketing manager,

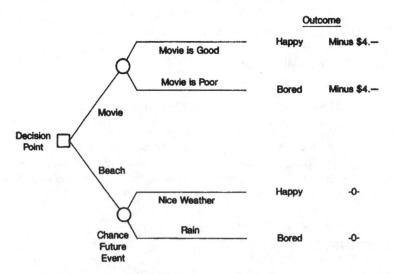

FIGURE 21-1. *A decision tree.*

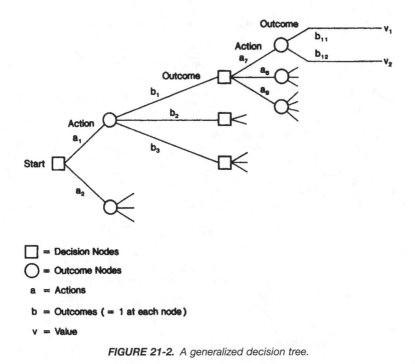

FIGURE 21-2. *A generalized decision tree.*

you learn that your company and two others were both solicited suddenly, and you all have an equal opportunity of winning the job. Thus, you appear to have one chance in three of winning $300,000 if you write the proposal.

After discussing this, you and the marketing manager realize that you have the opportunity to construct a working model for $45,000, and doing so increases the odds of your winning to fifty-fifty. The decision with which you are confronted is whether to build a working model (at a cost of $45,000) to increase the odds of winning a $300,000

Decision trees help you to structure your choices in the face of uncertainty.

before-tax profit from one in three to one in two. Figure 21-3 shows the decision tree and the problem analysis for this solution. Figure 21-4 is an alternative representation. In both cases, the branch in which the model is built has a higher value than the branch in which the model is not built. Therefore, your decision would be to build the model because it has a higher expected value. Note that this does not guarantee that you will win the job. Rather, it gives you a higher expected dollar value. If you use decision trees in enough cases, you will do better over the long run. But you might very well develop the model and still lose the job, thus ending up losing $45,000. (Note that, for simplicity, we have ignored the cost of writing the proposal and the time value of money in both situations.)

Imagine that you have constructed the preceding decision tree (Figures 21-3 and 21-4), but you and the marketing manager continue to discuss the situation prior

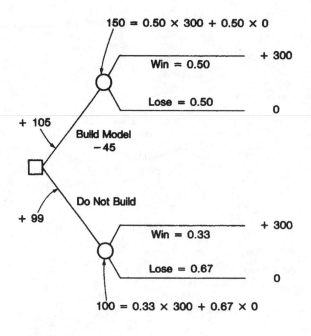

150 = 0.50 × 300 + 0.50 × 0

Win = 0.50   + 300

Lose = 0.50   0

+ 105

Build Model
−45

Do Not Build

+ 99

Win = 0.33   + 300

Lose = 0.67   0

100 = 0.33 × 300 + 0.67 × 0

Methodology:
- Construct diagram from left to right.
- Insert $ values from right to left.

FIGURE 21-3. *Decision tree for illustrative example (thousands of dollars omitted).*

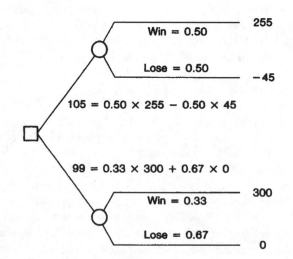

Win = 0.50   255

Lose = 0.50   −45

105 = 0.50 × 255 − 0.50 × 45

99 = 0.33 × 300 + 0.67 × 0

Win = 0.33   300

Lose = 0.67   0

FIGURE 21-4. *Alternative representation of Figure 21-3 decision tree.*

to building the model. You realize that whether or not you build the model, if you lose the proposal, you will have an opportunity to write a subsequent proposal to the winning contractor. Such a subsequent proposal will cost you an additional $10,000, but if you write it after losing, you will have an 80 percent chance to win a subcontract (worth $100,000 in before-tax profit) from the winning contractor. In this situation, you would decide not to build the model, as illustrated in Figure 21-5. (Another way to look at this revised situation is to realize that losing the original job is less serious, having an expected value of $70,000 instead of zero.)

Sometimes you are confronted with decision choices, for example, how to remove a resource constraint on two projects (as indicated in Figures 10-4 and 10-5), for which a qualitative decision tree (Figure 21-6) is useful. In this, you may be unable to quantitatively estimate outcome values, odds of occurrence, or costs of actions. However, you may be able to clarify the alternatives to the point where a better choice is possible.

MATRIX ARRAY

Quantitative

In situations where decision trees are not practical or are unwieldy for the analysis and comparison of alternatives, the matrix array may be a satisfactory aid. Figure 21-7 illustrates a quantitative weighting (or scoring) array. The key considerations in the particular problem are listed on

A matrix must list all significant criteria.

the left margin. In a computer programming project, the performance criteria might include processing speed, size of memory required, or similar considerations. For an airborne piece of equipment, performance issues might be weight, size, and reliability. In addition to performance targets criteria, the listing should include the schedule and cost implications of adopting that particular solution.

Next, weighting factors (the sum of which is equal to one) are attached to each criterion (or target). Then, a percentage, indicating the degree to which the solution satisfies the particular criterion target, is entered for each solution. Finally, each percentage for each solution in the body of the matrix is multiplied by the corresponding weighting factors, and the result is entered at the bottom of the solution column (e.g., for solution approach 2, $0.20 \times 90\% + 0.15 \times 70\% + 0.15 \times 50\% + 0.20 \times 70\% + 0.30 \times 80\% = 74\%$). There are four problems with this approach:

1. Deciding on weighting factors may be difficult, and a strong-minded person easily may argue for or manipulate these to favor his or her preferred option.
2. Choosing percentages may be difficult, especially for the schedule and cost targets, where the solution approach may exceed plan.
3. The highest ranking or weighted percentage value still may be inadequate.

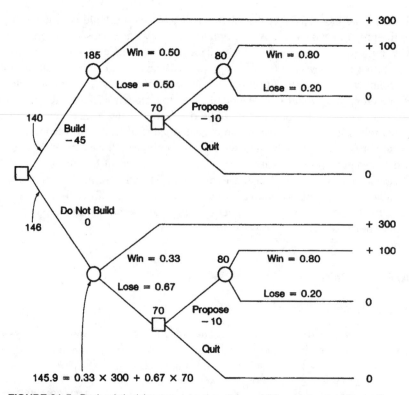

FIGURE 21-5. Revised decision tree to account for additional later decision options.

4. It does not consider people, in particular, the possibility that one solution approach is championed by an ambitious person who will work nights and weekends to accomplish it.

Computer software is available to mechanize this kind of approach, although a spreadsheet seems to be perfectly adequate. One form of software—the so-called fuzzy logic decision matrix—is also useful. The analytic hierarchy process is one commercial embodiment. In distinction to the purely quantitative matrix, this approach replaces the need for choosing weighting factors with the need to choose relative importance ratings. The potential advantage is that this is less prone to advocacy that can favor a forceful person's preferred choice.

Figure 21-8 is an example of a fuzzy logic decision using data from the matrix of Figure 21-7. The weighting factors from Figure 21-7 have been replaced by paired comparisons of criteria using the numerical scale of Table 21-1. As an example, two lines below "RATINGS OF THE CRITERIA" in Figure 21-8, you see a 1 and a 4. This means that criterion 1 of the pair on the previous line ("Performance Target A") is chosen as having somewhere between a "strong" and a "weak" importance over criterion 2 ("Performance Target B"). In the last pair of lines in that section of Figure 21-8, the "Cost Target" criterion has been chosen

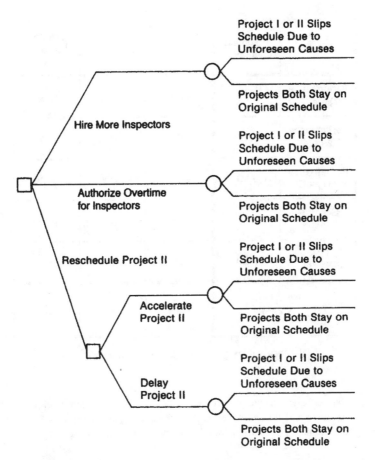

FIGURE 21-6. *Qualitative decision tree for resource allocation.*

as having a strong importance over the "Schedule Target." As shown in Figure 21-8, these data indicate that "Solution Approach 1" is preferred. Because the computational algorithm is obscure, it is harder for assertive persons to manipulate fuzzy logic to favor the result they desire.

Qualitative

Because of the manipulation to which the quantitative approach is subjected, it is often better to use a slightly more qualitative matrix. Such a matrix can explicitly consider people issues, especially where there are advocates for or opponents to a particular solution approach. Consider the following problem a project manager called to our attention:

> An employee volunteered to perform field work at a remote offsite location on my project and accepted relocation to that site for a one-year tour of duty. After two

Criteria	Weighting Factor	Solution Approach 1	Solution Approach 2	Solution Approach 3
Performance Target A	0.20	60%	90%	80%
Performance Target B	0.15	90%	70%	70%
Performance Target C	0.15	90%	50%	90%
Schedule Target	0.20	70%	70%	90%
Cost Target	0.30	90%	80%	70%
Weighted Percentage Value		80%	74%	79%

FIGURE 21-7. Quantitative decision matrix.

months, the field site manager reports that the employee is not getting along with the rest of the crew, and the employee calls and says that he wants to return. At this point, there are no other qualified employees who are willing to relocate.

In this kind of situation, there are really no tractable, quantitative approaches. In this kind of situation, a qualitative approach is all that can be used.

A qualitative matrix can use quantitative data where these are available (Figure 21-9). Note three aspects of this approach:

1. Numbers are used whenever possible.
2. People are given explicit consideration.
3. The summary identifies both favorable and unfavorable issues.

In addition, this kind of matrix can be used always, even early in the decision analysis, before all relevant numerical data are available.

Figure 21-10 shows a qualitative matrix array for the problem of selecting a three-axis "frictionless" support for an airborne instrument to allow 5 degrees of movement in three directions. The original perceptions of the factors to be considered are typed in the left margin. The first two solution approaches are listed. Both have many drawbacks. In Figure 21-11, a third alternative solution is identified, pointing out the need to consider one more factor (outgassing). Figure 21-12 shows

RATINGS OF THE ALTERNATIVES
.6 .9 .8
.9 .7 .7
.9 .5 .9
.7 .7 .9
.9 .8 .7

RATINGS OF THE CRITERIA

(1) Performance Target A	—	(2) Performance Target B
1 4		

(1) Performance Target A	—	(2) Performance Target C
1 4		

(1) Performance Target B	—	(2) Performance Target C
1 1		

(1) Performance Target A	—	(2) Schedule Target
1 1		

(1) Performance Target B	—	(2) Schedule Target
2 4		

(1) Performance Target C	—	(2) Schedule Target
2 4		

(1) Performance Target A	—	(2) Cost Target
2 5		

(1) Performance Target B	—	(2) Cost Target
2 7		

(1) Performance Target C	—	(2) Cost Target
2 7		

(1) Schedule Target	—	(2) Cost Target
2 5		

FDM PROGRAM OUTPUT
****

EIGENVALUE = 5.084712

EIGENVECTOR...
.1639196 5.148857E-02 5.148857E-02 .1639196 .5691838

ALPHA-VECTOR...
.8195978 .2574428 .2574428 .8195978 2.845919

CONSISTENCY OF THE PAIRED-COMPARISON MATRIX = .1029029
(SHOULD BE LESS THAN 1.0)

WEIGHTED FUZZY SETS...
.6579203   .9172701   .8328615
.9732402   .9122662   .9122662
.9732402   .8365695   .9732402
.7465221   .7465221   .9172701
.7409311   .5299099   .3623778

DECISION VALUES...
Solution Approach 1 - .6579203
Solution Approach 2 - .5299099
Solution Approach 3 - .3623778

Solution Approach 1 IS THE BEST CHOICE ACCORDING TO THE DATA YOU HAVE ENTERED.

FIGURE 21-8. One example of output from a fuzzy logic decision matrix software program.

TABLE 21-1 Explanation of Numerical Scale Used for Ratings of Criteria in Figure 21-8

Numerical Rating	Meaning
1-	Equal importance
2	
3-	Weak importance of one over the other
4	
5-	Strong importance of one over the other
6	
7-	Demonstrated importance of one over the other
8	
9-	Absolute importance of one over the other

a reasonably adequate solution for which the schedule must be studied further or perhaps accelerated.

Be honest about solution shortfalls.

To use this kind of matrix, list each solution across the top. You must do this conscientiously. When you write down the first solution and it clearly falls short with regard to one or more of the key criteria, seek additional solutions

Factor ＼ Solution	1	2	3	4	and so on . ..
Alpha					
Beta					
Gamma					
Delta					
Schedule					
Cost					
People					
Favorable					
Unfavorable					

FIGURE 21-9. Qualitative matrix.

Approach / Consideration	Conventional Outer-Ring Gimbal	Internal Gimbal				
Weight (Lbs)	50 ?	20 ?				
Size	Huge	Should Be OK				
Reliability	Bearings	Bearings				
Coercion (In-Lbs)	1 ?	1 ?				
Schedule	Quick	Less Quick				
Cost	Low	Higher				
People						
Pro						
Con	Weight, Size, Coercion	Weight, Coercion				

FIGURE 21-10. Qualitative matrix display.

designed to overcome the shortfall of the ones presently conceived. Usually, after you have written down two, three, or four solutions, each of which has one or more aspects of shortfall, you will identify some hybrid or new variant that comes close to satisfying all the key considerations identified. Even if the matrix array does not lead to a hybrid that clearly satisfies all significant considerations, using the array will clarify available tradeoffs and options.

PROBLEM-SOLVING STYLES

There are several styles of managerial behavior in working with others to solve problems. They include the following four:

1. *Command*—issuing orders
2. *Negotiative*—different groups evaluating common interests to reach agreement
3. *Collegial*—peers reaching decision by consensus

Approach / Consideration	Conventional Outer-Ring Gimbal	Internal Gimbal	Gas Bearing			
Weight *(Lbs)*	50 ?	20?	10-20			
Size	Huge	Should Be OK	OK, But Also Has Pump			
Reliability	Bearings	Bearings	Pump			
Coercion *(In-Lbs)*	1 ?	1 ?	Nil			
Outgassing	Oil	Oil	Support Gas			
Schedule	Quick	Less Quick	8-12 Months			
Cost	Low	Higher	Still Higher			
People			Jack			
Pro			Coercion			
Con	Weight, Size, Coercion	Weight, Coercion	Weight, Gas, Reliability, Time			

FIGURE 21-11. Qualitative matrix display with a third alternative solution identified.

4. *Advisory*—exchanging information and making subsequent decisions

If the information to solve the problem is readily available, a solution-centered approach to problem solving may be adopted. If the managerial authority is accepted and the manager has the information and the ability to solve the problem, a directive approach is appropriate, and a command meeting, one in which the manager issues orders, is the most appropriate meeting style. This does not mean that a different meeting style cannot be used, but a command meeting most likely will produce an effective solution under these conditions. The negotiative approach to problem solving is bargaining with different objectives but common interests, and in this case, a negotiative meeting is most likely to be effective. The prescriptive approach to problem solving is one in which a solution is solicited, and the presented answer may be only a tentative trial. For this, a negotiative or collegiate meeting style is most useful.

In the consultative approach, the parties use information sharing to diagnose the problem. In the reflective approach, which is useful if the problem is unclear,

Consideration \ Approach	Conventional Outer-Ring Gimbal	Internal Gimbal	Gas Bearing	2 Horizontal Flexures Vertical Mercury	3 Equal (35.2°) Flexures	1 Horizontal + 2 45° Flexures
Weight (Lbs)	50?	20?	10-20	~10	5-10	5-10
Size	Huge	Should Be OK	OK, But Also Has Pump	Reasonable	Small	Small
Reliability	Bearings	Bearings	Pump	Mercury Spillage	∞	∞
Coercion (In-Lbs)	1?	1?	Nil	~0.01	~0.01	~0.01
Outgassing	Oil	Oil	Support Gas	Mercury Vapor	Nil	Nil
Schedule	Quick	Less Quick	8-12 Months	3-4 Months	5-6 Months	4-5 Months
Cost	Low	Higher	Still Higher	~$15K	~$25K	~$20K
People			Jack	Hy	Mike	Mike
Pro			Coercion	Time & Cost, Coercion	Weight, Size, Reliability, Coercion	Weight, Size, Reliability, Coercion
Con	Weight, Size, Coercion	Weight, Coercion	Weight, Gas, Reliability, Time	Mercury	Harder to Make Parts	Time = ?

FIGURE 21-12. Qualitative matrix array with one reasonably adequate solution.

nonjudgmental restatements are acceptable. In both these approaches, the advisory meeting is most likely to be productive.

TYPICAL PROBLEMS

In most project management problem-solving situations, it takes considerable time and energy to find *an optimal* answer; thus the parties will accept a reasonable (or least objectionable) answer that they can all support. It is thus a matter of judgment about when to choose among the identified solutions and when to keep looking for more, better solutions. Honest people will differ (as they will in their perception of the problem and their evaluation of solution alternatives), and this must be both expected and tolerated. Use of a qualitative matrix or a purely qualitative decision tree can help to bring people's "mental models" to the surface so that a better understanding is created.

HIGHLIGHTS

- A seven-step approach to problem solving involves identifying the problem, collecting the data, devising a solution, searching for alternative solutions, adopting the best solution, implementing the solution, and auditing the outcome.
- Decision trees help you to evaluate assumptions and risks and choose the best course of action.
- Matrices, another aid to alternative selection, may be qualitative or quantitative.

Part 5

_____

Completing a Project

22

Closing the Contract

The fifth and last of the managerial functions discussed in this book is project completion. There are two processes involved in closing a project. We discuss contract closure in this chapter and cover administrative closure in the next chapter.

WINDING DOWN THE PROJECT

In Chapter 5, we introduced the topic of project life cycles, which is the series of phases that help the project team progressively elaborate work activities. Project management professionals frequently refer to the activities associated with the later project phases as *winding down*. Project success is enhanced by a thoughtful approach to winding down. During the last few months of a project, weekly reviews may be required, and during the last few weeks, daily reviews may be required.

Because there are many possible completion points and delivery conditions for a project, it is necessary to think through requirements and completion issues. Completion of an Apollo lunar project would differ depending on whether people defined it as the launch to the moon, the manned landing, the astronaut recovery, or completion of rock analysis.

The progressive elaboration of projects has implications for the illustrated personnel headcount and means that the kinds of personnel used in different phases will change. For instance, creative designers, very useful in the early phases, easily can become an obstacle to completion if they start gold plating the design instead of supporting implementation of the design. Figure 22-1 shows a simple generic three-phase project life cycle, which illustrates the point that project activities change during the duration of any project.

> **Personnel needs will change throughout the project life cycle.**

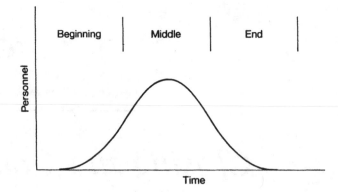

	Beginning	Middle	End
Construction	Site	Erection	Landscaping
Product Development	R&D, Market Research	Engineering and Manufacturing	Product Introduction
Aerospace System	Engineering	Assembly	Customer Test and Sign-off
Computer Software	Definition and Design	Programming	System Test and Acceptance

FIGURE 22-1. Project life cycle.

ACCEPTANCE

One important measure of project success is the customer's acceptance of the result. How do you do this? By obtaining and documenting the customer's verification that the project has delivered a result that meets the requirements.

> *Clear acceptance criteria make it easier to complete a project.*

In Chapter 2, we explained that meeting the customer's requirements is one measure of project completion and success. We provided examples of completion criteria, functional requirements, and performance requirements. We also pointed out some of the difficulties that arise when people use ambiguous wording. For example, the customer could interpret the word *appropriate* differently from the project's quality control engineer. Verification criteria that seem clear and simple at the beginning of a project have a way of becoming contentious toward the end. Objectively measurable criteria for completion (such as relocate "this" facility and have it ready to operate by next week or test these two amplifiers for gain using a specified method before 9:30 A.M.) are best.

As we have emphasized throughout this book, an important project management responsibility is managing expectations. This is true whether the project is for an internal or external customer. Requirements issues are elaborated progressively throughout the project life cycle. You want to ensure that you have a valid requirements set that "truthfully" captures the real need and provides the boundaries so that the designers can develop a design that satisfies the requirements. If you have achieved this, you will deliver a product that really solves the user's problem. However, in

> *A valid requirement is one that correctly expresses the customer's need in a "solution free" manner that allows the project team to solve the "real" problem.*

most projects, the rush to get into design and execution means that people will shortchange the requirements. Often the situation occurs where the project built the product according to the requirements, but the customer is unhappy because the problem was not really solved. In this case, the too-common fault is that the customer specified the product design rather than described its requirements in a "solution free" manner, in effect micromanaging the project.

While performing the process of acceptance, be sure to review the original agreements on requirements and scope and modifications to them. Be certain that completion criteria are stated clearly in whatever documentation initiates your project, preferably quantitatively and ideally with unambiguous means to determine how compliance will be determined. Further, experience shows that customer dissatisfaction is most frequently associated with missing performance requirements rather than missing functional requirements. You should pay particular attention to performance requirements.

Responsibility for delivered goods after they leave the contractor's facility is always an issue to be addressed in the requirements. Consider the situation where the customer has witnessed final system tests at your company and accepts the system. After delivery to the customer's plant, the system does not perform adequately. (In the case of the Hubble Space Telescope, a performance inadequacy was not discovered until the telescope was tested in orbit, which was years after the primary mirror left the optical manufacturing plant.) What can the project manager do? At a minimum, you are going to have to participate in fact finding. At one extreme, the acceptance criteria will have been absolutely clear; the tests at your company will have been unambiguous; participation in and witnessing of these tests will have been done by qualified and responsible senior customer personnel; and the performance shortfall at the customer's facility will be attributable entirely to the customer's actions.

More likely, questions will arise about the acceptance test, the specific equipment used to perform it, or what might have occurred to change the system during delivery. Regardless of contract form or your desire to obtain more business from that customer, the customer's final happiness with how well the system or product solves his or her problem will affect your reputation, so you may have to engage in a lot of extra, perhaps unpaid, effort to help get the system or product working. In some cases, you may be able to show that the fault was the custonmer's and get him or her to pay you for your extra effort.

Once the contractor has turned over possession of the goods, the contractor's control is greatly diminished. Therefore, it is important to have clearly stipulated conditions for acceptance, including payment terms.

Some projects require postcompletion service and support.

Continuing service and support may be part of the requirements. If so, it must be understood who is to pay for them and when. This is sometimes negotiated as part of the contract closeout. The contractor should view continuing service and support as an opportunity and not merely as an obligation. His or her employees will be working with the customer's personnel, providing continuing service and support, if it is included in the project. In so doing, they will have informal opportunities to explore ideas with the customer's personnel and hear about real problems the customer is facing. Thus, these contacts provide the basis for future business opportunities.

MANAGING SCOPE CHANGE

Completion requires objective and measurable criteria be attained, which ideally solves the customer's problems.

As discussed in Chapter 20, you should anticipate changes of scope, and those changes will affect the acceptance criteria. It is natural for the customer to expand his or her requests for more functionality and performance. Uncontrolled feature creep ("better" is the enemy of "good enough") is insidious, but a few well-understood and agreed-to goal changes, as depicted by shaded circles in Figure 22-2, can be tolerated.

DOCUMENTATION

Most requirements include the delivery of documentation (deliverables), as well as some other tangible output (results). Such documentation, in addition to a final report, might include a spare parts list, instruction manuals, and as-built drawings. Often, customers will specify the format of deliverables in their requirements.

Be sure to have a plan and schedule to complete all required project documentation.

Because furnishing these items may delay completion of the principal project output, they should be identified as a separate line item in the original contract. Thus, billing for the principal portion of the contract may accompany its delivery, and billing for the documentation reports may follow later. Figure 22-3 illustrates a conservative time plan to ensure that final documentation reports are delivered prior to the stipulated deadline. An analogous schedule could be constructed for hardware (although the shipment method might require use of a heavy-duty freight carrier).

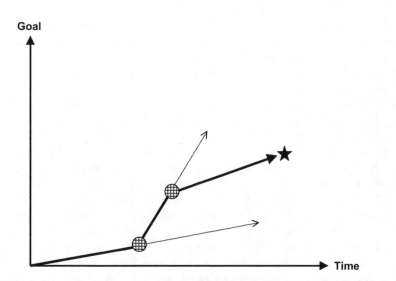

FIGURE 22-2. *Scope control requires that necessary goal changes be agreed to and clearly defined.*

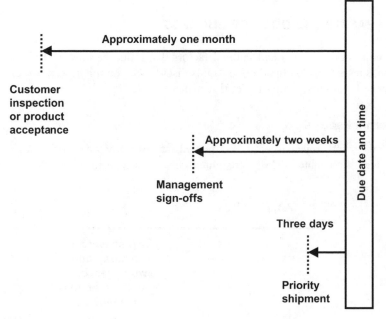

FIGURE 22-3. *A hypothetical, conservative completion schedule to ensure on-time delivery of required documentation.*

It is frequently difficult to complete documentation at the end of a project for two reasons. First, many technical specialists are poor writers or are reluctant to write. Second, in many instances, the people who have essential knowledge have long since been assigned to some other activity and are no longer working on the project.

The second problem can be eliminated and the first problem diminished by what we call *incremental documentation.* At the project's inception, you can prepare an outline of all final documentation and include this in your definition (or proposal) statement of work. Throughout the project, the responsible task manager or other specialist provides a few sentences or paragraphs when each key task is completed. Figure 22-4 illustrates this concept of incremental documentation, for which each task (the solid line) has a brief documentation appendage (the dotted line continuing after the solid line) during which the task manager completes a few sentences, paragraphs, or pages pertinent to the task just completed. These are inserted into the outline at the correct place. This is relatively painless and provides the essential information for final documentation later. These contributions might be provided via a local area network or on computer disks, thus reducing transcription requirements. Figure 22-5 illustrates how incremental documentation might be shown using a linked Gantt chart. Four documentation tasks (A*, B*, C*, and D*) have been inserted to follow the respective tasks, and all are shown as additional predecessors to the final report. (Each of the documentation tasks is shown as having very brief expected time durations.)

INCREASING THE ODDS OF SUCCESS

There are both external and internal factors that influence how well a project accomplishes its results. In addition to this discussion, you might want to review the Chapter 11 discussion on project risk management.

External Factors

Customers for projects seem to be divided into two broad categories: knowledgeable and shortsighted. The shortsighted customers tend to emphasize the buyer-

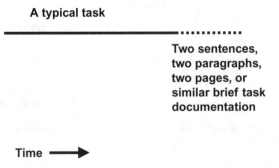

FIGURE 22-4. The concept of incremental documentation.

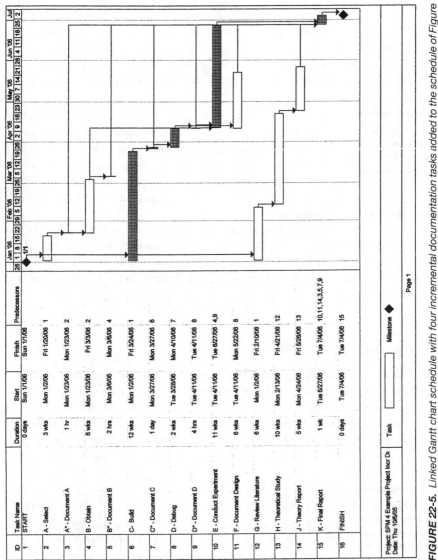

FIGURE 22-5. Linked Gantt chart schedule with four incremental documentation tasks added to the schedule of Figure 7-18.

versus-seller relationship and to some extent create an adversary relationship between the two organizations. Conversely, knowledgeable customers realize that their stake in project success is ultimately just as great as that of the performing contract organization. Thus, a knowledgeable customer will become involved in the project in an effective, as opposed to destructive, manner. Such a customer will specify expected reviews and include them in the original job definition. Beyond this, he or she will attempt to ask the tough questions and to carry out probing reviews of the contractor's work, not to embarrass but to ensure that all significant issues have been dealt with appropriately. Any required changes will be negotiated intelligently.

A knowledgeable customer, high priority, and clear objectives aid successful completion.

High-priority projects inevitably seem to have better outcomes than lower-priority projects because they tend to win all competitions for physical and human resources. This is not to say that low-priority projects lack top management support; top management clearly wants all projects to succeed, but the lower-priority projects are at a relative disadvantage.

Clear and stable project objectives are a *sine qua non* of project success. Objectives can and do change during the course of many projects, but not on a daily or hourly basis. Thus, committing these objectives to writing helps to fix them in everyone's mind. Revising them when they must occasionally change is also a requirement of success.

Internal Factors

A qualified, experienced, competent leader is vital, as is a balanced team. Having a team with a balance of skills and getting teamwork from it are essential. People with very similar backgrounds tend to get along better, so it is easier to promote cooperative behaviors in a group composed entirely of, for instance, electrical engineers. Nevertheless, a successful project usually requires a team composed of more than electrical engineers. Thus, the project manager must build an effective and harmonious team.

A good leader, a balanced team, the right-sized work packages, careful replanning, and orderly termination contribute to project success.

Having the properly sized work packages helps you to avoid two potential problems. Complex, difficult work packages should not be assigned to unqualified people, who may be overwhelmed by them. Simple work packages should not be assigned to senior people, who will not be challenged by them.

Because projects deviate from original intentions, planning is a continuing project activity. Project wrap-up and completion, especially the reassignment of personnel, require active planning well.

TYPICAL PROBLEMS

> Sometimes not only subordinate personnel but also the project manager must change during a project's life. The manager for the initial phases may be great at the inception but become stale with time or bored by routine wrap-up activities. The solution in this case is to change project managers, and both upper management and the project manager must be alert to this possibility, which then requires time for the replacement project manager to review (and perhaps revise) the plan so that he or she is comfortable with it.

HIGHLIGHTS

- The satisfactory completion of a project depends on requirements and other agreements reached, as modified by the organization's change control system.
- Personnel needs may change throughout the project life cycle.
- The project manager must realize that project completion may not be good for all involved parties and plan for an orderly end well in advance of its scheduled time.
- Both internal and external factors contribute to project success.

23

Final Wrap-Up

In the preceding chapter, we discussed project completion in terms of meeting the agreed-on customer requirements. In this chapter, we discuss the administrative closure activities that provide for the final wrap-up.

In the preceding chapter, we covered the contractual closure of a project where we confirm that we have satisfied the customer's requirements. Final wrap-up, or administrative closure, involves a final set of project activities.

Not everyone has the same stake in completion, and the project manager must understand the differences.

Project completion may be viewed as a boon or doom. The customer, the contracting organization, the project manager, and the project personnel may not all see it the same way. Thus, there is no reason to assume that all four parties will have the same view of project completion. Project managers must realize that they may have a very different stake in ending the project than the other three parties.

PEOPLE ISSUES

Administratively, people need to be "released" from the project to their "home" department, reassigned to another project, or possibly have their employment terminated. Increasingly, organizations have the project manager review the performance of the people working on the project team, and this project-level performance review is factored into the individual's overall contribution to the organization. It is also a good idea to send a personal letters of thanks, appreciation, or praise to the individual contributor and copy his or her boss. Project completion requires reassignment of people. We have now come full circle. The (temporary) project is no longer imposed on the rest of the (permanent) organization. This frequently will necessitate a reorganization of the parent entity because the mix of remaining project work is such that the previously satisfactory organization is no longer appropriate.

The other crucial aspect of personnel reassignment is timing. If a person's next assignment is a choice one, he or she normally will be anxious to start and will lose interest in completing the present project. Conversely, if someone's next project assignment is undesirable, he or she may stall. When no assignment is obvious and layoff or termination is probable, personnel even may attempt sabotage to stretch out the present project assignment. Incentive bonuses for timely completion may be useful to counteract this potential problem.

A person's perception of what will happen when the project ends will affect his or her work as termination approaches.

The project manager can cope with these tendencies to some extent by selecting the time that he or she informs project personnel of their next assignment. However, if the contracting organization has a reputation for terminating personnel at the end of projects, there is little a project manager can do. The best situation is one in which all project personnel can count on their good work being recognized and appreciated and there being a selection of future assignments.

Even if no specific new project assignment is available when personnel need reassignment, there are still options. For example, personnel can write an unsolicited proposal, prepare an article for publication, work on an in-house development effort, or attend a short course or seminar. Temporary assignments such as these can be used constructively to fill in valleys in the project workload. They also can be used as a motivational tool if they are authorized to make participation a mark of recognition for a job well done.

There are many options for personnel reassignment when the project ends.

People have the "free will" to pursue other employment opportunities. However, "no compete" clauses in recruiting customer personnel into the contractor organization or vice versa are often in the initial contract. If they were not, the customer might hire the contractor's personnel to perform the continuing service and support the contracting organization, presuming that it was going to perform and be reimbursed for.

In addition, as the project manager, take this as an opportunity to send a brief wrap-up report to management and to highlight some of your contributions to the success of the project. For the project manager, completion may be an opportunity for promotion, but many project personnel may find themselves looking for a new assignment and even a new employer. If the project was managed badly, its manager may receive a less favorable assignment in the future, and personnel who did an outstanding job may have a choice future assignment.

LESSONS LEARNED AND AUDITS

When any project ends, for whatever reason, the organization is altered. There is a new body of knowledge in the organization. This is not just tangible information but also new skills for many people. New working relationships have been established, both within and external to the organization, and these alter the informal organization, even if the formal organization is still unchanged.

Every project manager should organize a project postmortem. These are also known as *postcompletion* or *postimplementation evaluations* or *reviews* or, perhaps more helpfully, as *postproject learning.* Some of the aspects to investigate and draw lessons from are the procedures that were used; any processes, tools, or methods employed if hardware was involved; the abilities of the organization and individual staff members; and any training that would be desirable. The goal is to record the "lessons learned." As one experienced project manager observed, "Learn from others' mistakes; it takes too long to make them all over again."

Because it is difficult to be objective, the postmortem might be conducted by an outsider or a person not intimately involved in the project. He or she should interview each of the key participants and provide a summary report. This report might be presented in a briefing session for the key project personnel before its final submission to the sponsoring organization. Steven C. Wheelwright, a Harvard professor, once wisely commented, "Individuals learn; corporations don't."

His point is that an individual "once burned" will not be likely to repeat the same mistake on a subsequent project. The corporation, however, will not have that lesson unless it is both recorded and disseminated. The goal of such a report is to help the organization avoid unnecessary mistakes in the future and point out how to revise procedures so that similar future projects can be carried out more effectively and expeditiously. In some cases, it is best to prepare for such a postmortem by having the outsider who will conduct it interview key project personnel at two or three important points during the project to capture their thinking at those points while it is still fresh.

> *Those who cannot remember the past are condemned to repeat it.*

There often are postcompletion audits, especially in contracts performed for governmental entities. Such audits may occur many years after the project work is otherwise complete.

> *A project is not complete until the customer pays the bill.*

It is therefore essential to retain records for the required duration and even more important to file and document them in an organized and thorough way. Many of the people who could explain some audited item may no longer be available when the audit is performed, so the contractor might lose an important claim if this is not done. Some portion of the final payment on a contract may be withheld until the audit is completed. Therefore, a final financial summary may not be possible until the audit is performed, and this can be months or even years after the other work is finished.

What can you do if your customer does not pay the bill? The responsibility for obtaining payment usually resides with the accounts receivable department and thus is outside the normal purview of the project manager. Nevertheless, an effective project manager may be able to help persuade a slow-paying customer to pay promptly. In some cases, the customer may be withholding payment because he or she feels that the project has not delivered to the requirements, and usually the project manager can provide clarification to resolve this problem. Also, if the project manager wants further work from this customer, he or she does not want the accounts receivable department to take hostile action against a delinquent account. Thus, the project manager and the accounts receivable department must discuss these issues and mutually decide on the mix of persuasion, cajoling, and legal action to take.

INTELLECTUAL PROPERTY AND OTHER OWNERSHIP RIGHTS

Patents and special facilities required for contract performance have ownership value. In general, if it is a cost-reimbursable contract in which the customer pays all the incurred costs, ownership rights revert to the customer at the end of the job. If it is a fixed-price contract, ownership rights generally revert to the contractor unless otherwise stipulated in the initial contract. These can be points of contention unless they are discussed and clearly resolved in the initial contract. In any event, patent applications must be filed if any seem justified. The party of ultimate ownership must expect to pay for this activity.

Outline patent rights and hiring policies in the contract.

TYPICAL PROBLEMS

Basically, the problem here is simply doing it all. The press or excitement of new items to do often leads to the omission of some postcompletion activities. Organizations frequently never get around to performing a formal postmortem, so the same mistakes are repeated in subsequent projects. The solution is to recognize your responsibility in getting these things done.

HIGHLIGHTS

- Contracts often include assignments of patent rights and "no compete" personnel recruitment clauses.
- Records must be kept carefully in case there is a postcompletion audit.
- Managers should send letters of appreciation to project personnel and a wrap-up report to the boss.

Part 6

Other Issues in Project Management

24

Small Projects

Small projects have both advantages and disadvantages. Being smaller, they are easier to understand and less likely to get into difficulty. However, there is less time and money to recover from anything that goes wrong, and they inevitably lack high priority.

Like project tasks, small projects have a *relatively* smaller scope, budget, resource pool, and time. It is important to get both off to a good start by ensuring that you have proper resources in place, a clear idea of the customer requirements and project goals, and a clear indication of the priority of the project in terms of the enterprise's portfolio of projects.

SIMPLIFIED MANAGEMENT

Figure 24-1 is an illustration of a spreadsheet used to plan a small project. The overall plan for cumulative labor and nonlabor costs is also included. As actual costs are incurred, they can be entered and variances noted by altering the spreadsheet as shown in Figure 24-2. Note also

Small projects have a Triple Constraint to plan and monitor.

that the spreadsheets use weeks for the time horizon across the top, which may be a more useful time frame for a small project. This could be changed to months or any other convenient time frame. An alternative form, which lacks specific cost information, is illustrated in Figure 24-3. Clearly, if a spreadsheet is employed for planning and monitoring, additional columns and/or rows can be entered to include other data (e.g., names of specific resources, identification of predecessor and successor tasks, and so on) that are considered important for your small project. Remember, however, that you don't want to overwhelm an inherently small project with a mountain of paper and data.

You also can use computer-based project management software. If it or other computer software is not otherwise justified or required, forms such as those shown in Figures 24-1 and 24-2 may suffice.

	Week # =	1	2	3	4	5	6	7	8	9	10	11	12	13	14	15	16	17	18	19	20
Task A	Hours =	4																			
	Non-labor $ =																				
Task B	Hours =		6																		
	Non-labor $ =		50																		
Task C	Hours =			12																	
	Non-labor $ =																				
Task D	Hours =			6																	
	Non-labor $ =			50																	
Task E	Hours =			2																	
	Non-labor $ =																				
Task F	Hours =				8	16	16	12	10	8											
	Non-labor $ =					100	100	50	50												
Task G	Hours =									8	8										
	Non-labor $ =																				
Task H	Hours =											4									
	Non-labor $ =																				
Task I	Hours =											2	2								
	Non-labor $ =																				
Task J	Hours =													1		3		1			3
	Non-labor $ =															50					50
Cumulative Total	Hours =	4	10	30	38	54	70	82	92	108	116	122	124	125	125	128	128	129	129	129	132
	Non-labor $ =	0	50	100	100	200	300	350	400	400	400	400	400	400	400	450	450	450	450	450	500

FIGURE 24-1. A spreadsheet can be used as a simplified form for small project planning.

Week # =	1	2	3	4	5	6	7	8	9
Task P									
Planned Hours =				40	80	80	80	60	40
Non-labor $ =					1000	1000	500	750	
Actual Hours =				32	64	72			
Non-labor commitments =					1100	1200			
Hour Variance =				8	16	8			
Non-labor Variance =					-100	-200			

FIGURE 24-2. *Using a spreadsheet as a simplified form for monitoring (with actual charges included through week 6).*

PROBLEMS

There are four causes for the problems unique to small projects: tight schedules, tight budgets, small teams, and low priority.

Tight Schedules

A small project typically is planned to have a shorter schedule than a large project. Thus, the inevitable "getting up to speed" consumes a larger fraction of the available time. A one-week or one-month delay on a long program is less significant than the same delay on a two-, three-, or four-month program, a typical period for a small project. The implication is that small projects must be initiated at the very first opportunity, and the project manager must devote a relatively large amount of effort to ensuring schedule compliance.

Tight Budgets

A small project budget will be less than a large project budget. Hence, the absolute amount of money available for contingency must necessarily be less. When the inevitable revolting development occurs, there is less maneuvering room in which to cope with the consequences. The typical two-week lag in cost reports represents a significantly larger

Small projects lack time, money, personnel, and priority.

fraction of the overall project time and again leads to less reaction capability when some revolting development is discovered. In addition, the cost of time devoted to the program management function is likely to be a larger fraction of the total project budget. Thus, extremely close attention to cost is required on a weekly, if not on a daily, basis.

Small Teams

A large project typically can have the full-time attention of a functional expert (e.g., a thermodynamicist); a small project might have to accept the part-time as-

PROJECT TASK PLAN AND SCHEDULE

DATE _____
PREPARED BY _____ REV. BY _____
PAGE _____ OF _____

PROJECT NUMBER _____
PROJECT TITLE _____
WORK UNIT/DESCRIPTION _____

TASK NO.	TASK DESCRIPTION	RESP.	DEPENDENCY TASKS	HOURS		DATE		TASK SCHEDULE: PERIOD ENDING
				PLAN	TO DATE	PLAN	ACT.	YEAR

FIGURE 24-3. An alternative form for small project use.

Legend

△ SCHEDULE EVENT ONE TIME
△ RESCHEDULED EVENT NUMBER INDICATES RESCHEDULING SEQUENCE

▲ COMPLETED EVENT
△——— SCHEDULED EVENT TIME SPAN

▲——△ PROGRESS ALONG TIME SPAN
△——► CONTINUOUS ACTION

◇ ANTICIPATED SLIPPAGE
◆ ACTUAL SLIPPAGE (COMPLETED)

signment of such specialists. Thus, the small project must compete against other projects for the specialists' time. In some cases, especially where computer programming is called for, this can be a major problem. Each time a person begins a particular programming task, he or she will spend a certain amount of time "getting up to speed." Thus, time is lost in orientation. The problem may be worse with computer programming, but it is not confined to that specialty. Confronted with this reality, the effective project manager should attempt to bargain for full days whenever part-time resources are required.

Low Priority

High-priority projects will win each competition for any key resource. If you have a low-priority project and another person has a high-priority project and you both ask the model shop to make parts for your projects, the person with the high-priority project will have his or her needs satisfied first. Moreover, it is unlikely that a small project will ever have the same significance for an organization as a large project, which means that low priority is more common on small projects than on large projects. Finally, small projects have less visibility and therefore less chance for personal glory, so motivation can be less.

Imagine that you are spending half your time managing a small four-month project and spending the other half working on a much larger project. After two months, you discover that the small project is running late and will now require three-quarters of your time to complete on schedule. There are at least four options to consider:

1. Be late on the small project.
2. Request that you spend only one-quarter time on the large project that so you can spend three-quarters of your time on the small one.
3. Request paid overtime approval for the small project.
4. Work unpaid overtime on the small project.

The choice among these, and any other viable options, can be aided by a decision tree or a qualitative matrix, as discussed in Chapter 21.

TYPICAL PROBLEMS

There is another problem with projects starting small and growing in scope beyond the controls established for the small project. Again, "staying on top" helps, as does switching to more extensive, formal project management techniques as the project grows.

Watch for scope creep.

(continued)

Finally, the smaller team means that a small project is more likely to be adversely affected by a team member's illness, family emergency, or decision to resign. These potential events are outside the control of the project manager, and the best way to cope is to react very quickly.

HIGHLIGHTS

- Four causes of the problems unique to small projects are tight schedule, tight budgets, small teams, and low priority.
- Small projects have a tendency to grow in scope.

25

New Product Development Projects

New product development projects are common and important for many corporations. This chapter covers unique characteristics of these projects, provides a framework with which to understand how to obtain the desired profits quickly, and stresses the importance of ensuring that adequate resources are available.

WHY NEW PRODUCT DEVELOPMENT PROJECTS ARE UNIQUE

New product development (NPD) projects deserve special attention because of their strategic importance. NPD is an omnipresent economic undertaking of great commercial importance worldwide, and project management tools and techniques are especially useful for the varied projects that arise in NPD. Although the phrase *new product development* is used commonly, it embraces both tangible products (e.g., a new drug, child's toy, commercial airliner, or frozen-food entree) and less tangible services (e.g., a new type of certificate of deposit, combination of credit card terms, or frequent user program). NPD projects also provide a unified way to demonstrate the practical role of many topics previously covered in this book.

NPD projects differ in six important ways from other commonly encountered projects:

1. The Triple Constraint dimensions usually have particular characteristics:

    • Competitive pressures, the commercial "window of opportunity," and the substantial advantages of being the first market entrant with a new product are such that the schedule is highly visible and emphasized.

    Schedule is normally paramount for an NPD project.

- The performance goal is also important, but it is rarely an absolute standard. The performance goal normally is established as a marketing tradeoff with the forecasted schedule availability and relative benefits the product is expected to provide to the prospective users. In general, a simple new product can be developed quickly, whereas a more complex or higher-performing new product requires more development time. The performance goal also must include the target selling price and gross margin, so the expected *factory* cost is important.

- The actual *development* cost per se is relatively unimportant after the financial payback has been calculated and justified initially. The critical issue is more commonly the allocation of limited resources (principally human but also physical) among competing NPD investment opportunities. It is more important for senior executives to constantly ask whether the continued investment in a particular NPD project is the best (or, at least, justified) use of the firm's limited resources and not to focus on whether the NPD project is over- or underbudget.

2. They are self-funded by a company rather than undertaken as a contract to deliver some result for another organization. (NPD obviously can be contracted by a sponsoring company to a specialized NPD company, but this is not a common way to develop new products. Most NPD projects are carried out within the sponsoring company, although there may well be subcontracts to deliver parts, components, subassemblies, or production equipment.) Other things being equal, lower project cost will be more profitable than higher. The latter frequently results when projects are undertaken on a contract basis.

3. NPD can involve many different kinds of projects. These include products that are new to both your company and the market, new to the market (product-line extensions), and new to your company ("me too" products), minor variations on an existing product, sustaining efforts (typically technical assistance to the manufacturing operation), troubleshooting, and so on. These may be very small projects (a few person-hours or a few days) to very large (e.g., hundreds of person-years for a new commercial jet airliner). Some projects may require little technology (a revision to an operator's instruction manual); others may involve the complex integration of several technologies (e.g., mechanics, optics, electronics, software, and reagent chemistry for a new medical diagnostic system).

4. Isolated, so-called one-off NPD projects are common. Unfortunately, these are far less profitable than efforts that support an integrated family of new products based on a common platform. The latter provide numerous benefits to producers and can offer substantial benefits to buyers and users.

5. It is important to satisfy both a customer (the direct buyer) and a user (the person who actually employs the product). For many consumer goods, these are the same, but pet food and baby products are important categories for which the buyer and user are not the same. In most cases, there is also a distribution channel from the manufacturer to the point of sale, and the chan-

nel intermediaries (retail chains, stocking distributors, manufacturer's representatives, and others) are critical factors in commercial success or failure. Market research is therefore a critical ongoing activity during development.

6. Timely and cost-effective NPD inherently requires a multifunctional project team that carries out many tasks simultaneously. Coordination is therefore crucial and can be greatly encouraged by the use of a linked Gantt schedule (such as shown in Figure 13-2) where the nature of each function's involvement in each task is clearly indicated.

Multifunctional teamwork is crucial.

A GENERAL FRAMEWORK

Understanding some ways to quickly cash in on your investment in an NPD project can help. The next section frames the issue, exploring what profit means and why a lack of clarity may make it elusive, and then defines four intervals of time from inception to the end of the product's life.

The Profit Objective

The widely practiced stage gate or similar NPD process typically is portrayed to start with an idea (the proverbial "lightbulb") and end with "launch" and then a "pot of gold" or a "bag of cash," as shown in Figure 25-1. In fact, even if you have navigated successfully through "stages and gates," there is nothing automatic about (1) making enough money to justify the risky NPD effort or (2) making money quickly enough to yield a satisfactory discounted cash-flow return.

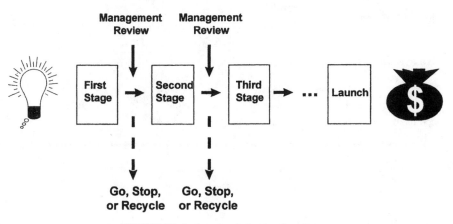

FIGURE 25-1. *A general stage-gate process.*

> **The profit objective is multidimensional.**

The profit objective is multidimensional and includes project-level and firm-level dimensions or components. There are often two simultaneous project-level measures, which may be comprised of some combination of market share, revenue, profit, competitive advantage, and customer satisfaction or acceptance. The firm-level objectives can include both a financial and a performance measure, such as return on investment, percent of profits from new products in recent years, the success-versus-failure ratio, and a fit-to-business strategy.

Regrettably, many firms have a structured NPD process, but the NPD project teams lack a clear understanding of either the firm's NPD strategy or the multidimensional profit objective. Lacking knowledge of the desired destination, an NPD project team depends on luck to be successful. In the following discussion, we aim to (1) broaden your view of NPD projects to encompass the entire process from inception to the realization of the profit objective (whatever that may be in your firm) and (2) identify a few ways to help you shorten the time from inception to profit.

Four Intervals (or Periods)

Figure 25-2 illustrates five events and four intervals, which provides a conceptual overview of an NPD project. In this figure, we use the abbreviations *FFE* (fuzzy front end), *S&G* (the traditional stages and gates portion of the process, which is sometimes mistakenly seen as the entire effort), *SS* (the postlaunch period of sustained sales until the time at which the profit objective is achieved), and *CS* (continued sales subsequent to achieving the profit objective). CS is "gravy" and will not be considered further in this chapter.

> **There are three general ways to shorten time to profit.**

Figure 25-3 shows that it is possible to shorten the time to profit by shortening the three initial intervals. Alternatively, it may be possible to lengthen the FFE to subsequently shorten S&G and/or SS or to lengthen S&G to shorten SS.

Five events

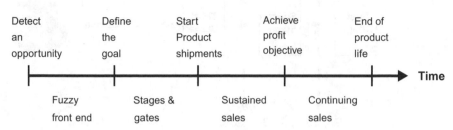

Four intervals (or periods)

FIGURE 25-2. Five events and four intervals (or periods) between the events.

Nominal plan

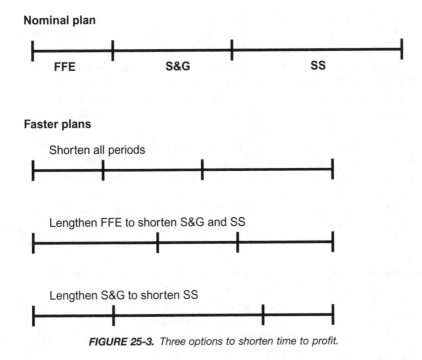

Faster plans

Shorten all periods

Lengthen FFE to shorten S&G and SS

Lengthen S&G to shorten SS

FIGURE 25-3. *Three options to shorten time to profit.*

Fuzzy Front End   Figure 25-4 shows that you can initiate the FFE with any of four stimuli. All four of these factors (strategy, ideas, technology, and markets) must be considered carefully and resolved satisfactorily in this initial interval. The goal of the formative period is to remove uncertainties, because a development schedule is not credible if there are major unknowns about the market, the technology, or the production process. Failure to carefully execute the FFE is often the cause of extra, more costly effort during the subsequent S&G or SS intervals.

Project management tools are most useful in situations where what has to be done is known with reasonable certainty. They are less useful for the overall dis-

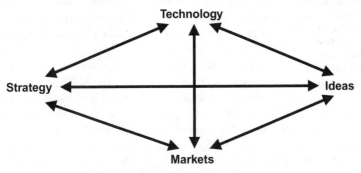

FIGURE 25-4. *Four FFE factors.*

covery research situation, where each experimental result can dictate a new course of action; however, the tools are invaluable for managing each experiment to increase the likelihood that each experiment—once decided on—is completed quickly and efficiently. Project management tools and techniques are also often helpful in managing market research conducted in conjunction with research experiments during the FFE. To put this differently, the FFE and fundamental discovery research should be structured to quickly run numerous knowledge-building experiments, which are often small projects.

As one example of a company that has developed a systematic approach to the FFE, Kodak previously started with an idea (generally from a technologist) and then went directly into its S&G development process. The company now selects an opportunity and then goes through its "discovery and innovation" process (which entails forming an innovation charter; seeding ideation; building models, samples, and prototypes for screening; and developing a business plan). Only then does the company initiate the S&G part of development. Other companies have similar approaches to NPD.

> **Strategy, ideas, technology, and markets must be considered in the FFE.**

The model shown in Figure 25-4 is general and therefore potentially very useful. For example, either market or technology discontinuities can provide excellent entry points into the FFE. Senior executives may want to change the firm's strategy (but only thoughtfully and carefully) as a result of a new "killer" idea.

The "window of opportunity" for any effort must be considered early in the FFE and may lead to an early termination or require the acceleration of an effort. In all cases, the end of the FFE is a solid business case that is judged against understood criteria.

Stages and Gates   Figure 25-5 shows the efforts in S&G (set specification, time-critical new product development, and continuous learning) and the FFE and SS intervals that, respectively, precede and follow. The reason to show these other activities is that many of the resources required for S&G are also involved, and your firm's resource allocation (or lack thereof) has important implications for the timeliness of S&G.

> **The time to market depends on the difficulty of the specification and the effectiveness of the resources.**

Figure 25-6 (which is based on the concept illustrated in Figure 8-5) depicts the key relationships that affect speed during the S&G interval. This is the critical S&G tradeoff. In general, a product with a more difficult specification normally has a longer time to market (i.e., launch) than one with an easier specification. The time to market is crucial in some situations, such as in the $1 billion or greater investment for a new generation of a dynamic random access memory chips, where the vast bulk of profits is earned in the first year of sales.

In Figure 25-6 and subsequent Figures 25-7, 25-8, and 25-9, we illustrate some of the reasons NPD schedules are longer than optimists (and some senior executives) expect. It is difficult to achieve the "best possible schedule" when appropriate

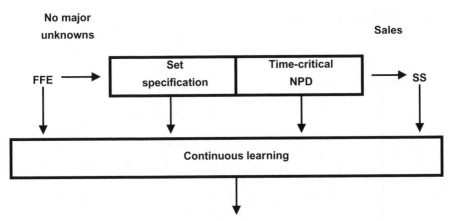

FIGURE 25-5. *Key S&G activities.*

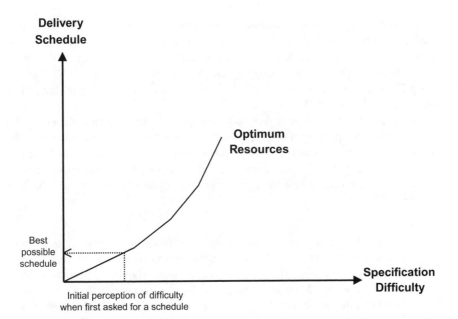

FIGURE 25-6. *Some reasons why project schedules are longer than what is the best possible schedule. See also Figures 2-3 and 8-5.*

Impact of Realistic Resources

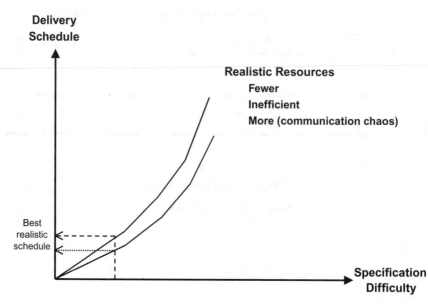

FIGURE 25-7. *Realistic resources are not ideal.*

resources are not available or assigned. As shown in Figure 25-7, realistic resources lead to a longer time schedule. A better understanding of the complexities of achieving the specification leads to a still longer schedule, as shown in Figure 25-8. Finally, in many cases, the design specification is altered during the NPD project to make it still more difficult, and the delivery schedule is further delayed, as illustrated in Figure 25-9.

However, at any level of design specification difficulty, the time to market can be shortened by applying more effective resources; the converse is also true, that use of less effective resources causes a longer time to market. Thus, to shorten the time to market, you must aim for simpler designs or plan to use more effective resources (either people or equipment). Here are three ideas to shorten the NPD interval:

1. *Clarify the requirements specification.* Specifications can involve both objective (quantitative) and subjective (qualitative) elements, as discussed in Chapter 2. Like any other project, requirements should be verifiable, and it takes time to decide on the appropriate verification criteria. In a small private company, the owner typically makes the decision, so he or she must be kept informed. In a large public company, it can be harder to identify a single

Impact of Understanding Specification

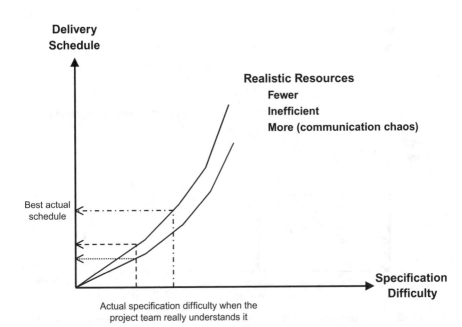

FIGURE 25-8. *The specification is more difficult than it first appeared when it is thoroughly understood.*

decision maker, but this is what you must try to do. Lacking this, developers end up working at cross-purposes, and time is lost.

2. *Capitalize on a product family.* Another way to simplify specifications is to plan to develop a product family (i.e., use planned releases to spread the related new products over a reasonable time span). This approach offers seven potential benefits:

You can start quickly with a relatively simple member.

Early entry provides real market data.

Lower costs are possible owing to common parts and processes.

There are fewer risks because portions of the product are already proven.

Follow-on members can reach the market quickly.

It is easier to defer "feature creep."

You can provide a "hook" to extend product life (especially for software upgrades).

There are numerous examples of product families in the consumer electronics market. Go into any mass merchandise outlet (such as Circuit

Better is the Enemy of Good Enough

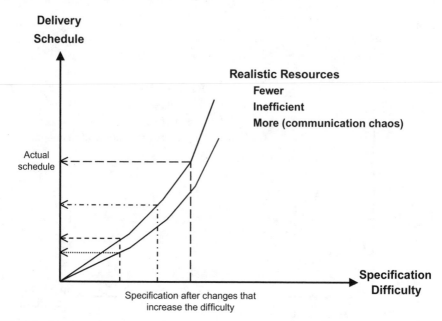

FIGURE 25-9. *The specification changes either at sponsor request or because workers strive for perfection.*

City, Office Depot, or similar), and look at the wide variety of products with slightly different features and prices being offered under a single brand (e.g., Sony tuners or CD players). Hewlett-Packard (and other PC vendors) has replicated this tactic with its ink-jet printers and its personal computer family. In the case of H-P's Pavilion PCs, the initial market introduction offered buyers seven models.

3. *Use flexible gates.* Flexible gates (Figure 25-10) in a stage-gate process can shorten the time to market during the S&G interval. However, replacing rigid gates with permeable and permissive gates increases risk somewhat. Nevertheless, on balance, using flexible gates is often the lesser evil.

Problems after market launch can delay achievement of the profit objective.

Sustained Sales Problems The list of problems that can plague a new product when it reaches the market is almost endless, and sadly, new types of problems continue to emerge to frustrate prospective users. No list, therefore, can be exhaustive, but the following common problems are illustrative:

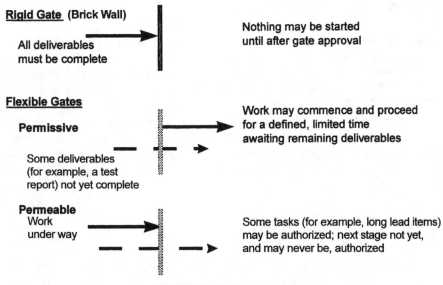

FIGURE 25-10. *Gate options.*

Users encounter unanticipated difficulty with the product, such as arcane instructions, application incompatibility, or inadequate performance.

The producer or, worse, the user is saddled with high maintenance costs.

The producer or someone in the supply chain is forced to maintain a large spare parts inventory.

The producer incurs an unexpected product warranty cost, and perhaps must cope with returned goods.

The producer discovers that the price-to-cost ratio is less than planned.

There are actions that can and should be taken in the FFE and S&G intervals to avoid SS problems and thus shorten the time to profit. Again, the list could be almost endless, so a few illustrations must suffice:

Design the product for ease of use.

Invest in really excellent user manuals.

Maintain a well-staffed help line.

Spend enough time to ensure that the design is robust and satisfies all advertised performance specifications.

Plan for rapid product distribution after launch.

Facilitate the efforts of the sales force by providing helpful selling tools, conducting an active promotion and advertising campaign, and otherwise stimulating their motivation.

RESOURCE OVERLOADING

Understand the availability of needed skills.

The key to ensuring a short time to profit is the proper allocation of a company's limited resources. Both resource capacity and flexibility are limited, no matter how large the organization. In the case of baseball, we can ask how many great hitters make great pitchers or how many catchers make great shortstops. In the world of NPD, only a few highly trained specialists are flexible generalists, so a mechanical engineer who is available may not have the skills needed for a particular NPD project. Furthermore, a critically required test facility may not have the capacity to simultaneously accommodate two or more development projects.

The Resource-Allocation Problem

In most companies, there are too many NPD projects under way at any given time, so all are starved for resources and move forward at less than the fastest speed possible. Worse yet, any problem with an existing product threatens immediate revenue and becomes a current crisis. This normally preempts development resources and thus interrupts some or all NPD efforts. Not uncommonly, longer-range development projects are sacrificed for those of shorter range.

Two Solutions to Consider

As we have stressed, an important project management responsibility is to manage expectations and generate realistic commitments of resources. A good starting point is understanding the organization's portfolio management strategy and working to align every project with the priorities established by the portfolio management strategy. To support top management's understanding of resource issues, you might want to take advantage of analyses generated by enterprise and project management software.

1. *Project priorities.* In Chapter 2, we discussed the growing presence and toolkit of project portfolio management that provides a framework for establishing project priorities and communicating them. Thus, the organization can apply sufficient development resources to the highest-priority effort. If the highest-priority project never has to wait for resources it can productively use, it can be completed faster. Thus, the resulting product can begin to earn income, and resources can be redeployed. These resources then can be applied to the second-priority project, and so on. If a company has enough resources to support several development efforts simultaneously, it can do so, but even in such a situation, the practice of concentrating resources on a

smaller number of efforts so that they move faster still provides an overwhelming advantage.

2. *Project management software.* As shown in Chapter 10, project management software can be used to identify resource conflicts before they occur. It also can be used to encourage multifunctional teamwork by serving as both a coordination and responsibility matrix. As illustrated in Figure 13-2, columns for each organizational function can be used for each task to indicate the nature of its responsibility in that task as accountable (A), concurrence required (C), or input required (I).

Resource-Allocation Strategy

Every company, even the very largest, has limited human and physical resources. Some of these resources must be devoted to managing the existing products and markets and dealing with the day-to-day distractions that consume time (e.g., staff meetings, holiday and other social events, personnel issues, and so on). Too frequently, NPD project resources consist solely of those that are left over. In most companies, however, these resources are not thoughtfully allocated among the following:

Minor product refinements, substantially new products, and major breakthrough efforts

The FFE, S&G, and SS intervals

High- and low-risk efforts

Near- and long-term projects

Resource allocation is an important operational issue for top executives: The use of project management software can elucidate specific potential resource shortages; thereafter, specific project priorities must be assigned to help resolve or overcome these shortages. Sadly, many executives do not recognize this need. As one NPD project manager said, "In what ways would it be best to communicate (and stress the need) to our upper management of the importance of establishing project priorities?" In other cases, the priority decision is left to the marketing department, but an NPD project manager complained, "How can we set up a project priority process in which marketing considers input from all functional groups to ensure that the priorities realistically fit the limited human resources?" As project manager, you must get top management to clarify whether or not your NPD project has the highest priority; if not, you must clearly inform the relevant executives about schedule consequences caused by resource limitations.

Top management must allocate resources to the most important NPD efforts.

TYPICAL PROBLEMS

The initial specification and development plan for an NPD project commonly results in a time to market that is longer than acceptable, as illustrated in Figure 25-11. It is easy to invoke arbitrary reductions of time, but these will not actually occur without a change of plan. Figure 25-11 is a disguised real case in which the initial time to market was about three years, but market competition required it be one year or less. In this case, the shorter goal was achieved by doing the following:

1. Putting several serial tasks in parallel.
2. Simplifying the design specification for an initial interim product release.
3. Finding and using more effective resources.
4. Using staged market introductions in a "family" plan of releases, with a very simple first product (P1) introduction accompanied by promised introduction dates for the two planned follow-on versions.

Premature promises by the members of the marketing or sales functions create another common problem for NPD project managers. In some cases, these functions optimistically assure prospects that certain specifications or delivery dates can be met. In the world of mass-market computer software, the word

(continued)

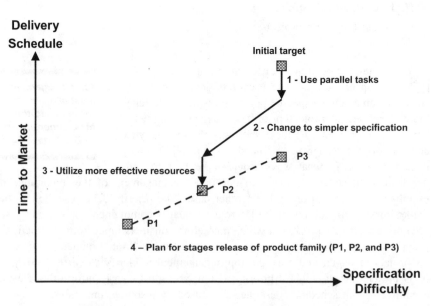

FIGURE 25-11. *An example of shortening time to market.*

"vaporware" has been coined to characterize early product announcements. In many cases, the claims have not been checked with the development or manufacturing functions. This can lead to a lack of credibility in the marketplace or to an excessive and stressful workload in the departments responsible for fulfilling these commitments.

HIGHLIGHTS

- Clarify and agree on the profit goals that are the objective.
- Manage all three initial intervals (FFE, S&G, and SS) to minimize the overall time to profit.
- Allocate limited resources to accelerate the highest-priority efforts.

26

Project Management Software

This book contains several examples and illustrations of the application of project management software to activities such as project scheduling and team coordination. Project management software is a tool. You need to manage the project personally and leverage the abilities that the tool offers.

Throughout this book, we have emphasized that project management entails strategic decisions and activities such as building teams, setting expectations, defining requirements, and recognizing risks.

WHEN AND WHERE TO USE COMPUTER PROJECT MANAGEMENT SOFTWARE

In Chapter 5, we wrote that critical thinking is essential and cautioned against making the common mistake of assuming that project planning is simply a matter of filling out a software template. We recommended using use a pencil-and-paper approach to project planning to help the team construct a project model that all can understand rather than simply focusing on the artifact of a Gantt chart or some other kind of document. Your goal is to drive the planning to a baseline "freeze" for execution and control.

A valid project model, automated with software, provides the project manager with a "reality test," which forces him or her to think through the project in sufficient detail to develop an optimal (or at least improved) way to complete it successfully. We have provided several examples and illustrations in this book of the output of computer project management software (Figures 7-17, 7-18, 7-19, 13-2, 17-5, and 22-5). This kind of software is constantly changing with the addition of ever more features and functionality and the concomitant complexity. You should understand what help such software can provide and take advantage of this where appropriate to you and your projects.

Increasingly, companies are deploying enterprise-wide project management software systems. However, you should assume that your key project personnel do not understand the output diagrams, reports, or formats. You should train these people in all this material if you wish them to use it effectively. The same is true for outside organizations.

Be prepared to train your key personnel in software use.

We have emphasized the importance of people to successful project management. As a tool, project management software can facilitate the team's communications so that the computer's outputs represent *their* collective thinking. One approach is to use an optical projector to exhibit the computer monitor's display on a large screen for team discussion during meetings of the key project personnel.

Use software appropriately as a communications and decision-support tool.

CAUTIONS WITH COMPUTER PROJECT MANAGEMENT SOFTWARE

First, as with all software, remember the aphorism, "garbage in, garbage out" (GIGO). Unless you take the time to enter data that are as accurate and meaningful as possible, your output may be worthless.

Second, learn the terms your software uses. There is good news in that over the past several years, the project management profession has worked to develop a generally recognized (standard) set of project management terms. The better commercially available software packages are adopting this standard language, but there are exceptions.

Third, avoid letting people sit at a computer terminal performing "what if" tradeoffs and updating the schedule instead of spending sufficient time interacting with the human resources on your project. If they spend excessive time isolated at their computers, the software will impede your (and their) ability to carry out the leading activity. Similarly, top management should not insist that an unqualified project manager use project management software with the expectation that the software will make him or her a good project manager.

People, not computers, are the basis of successful project management.

OTHER SOFTWARE

Chapter 21 discussed analytic hierarchy software to assist with decision making. There are many other kinds of software that can be helpful, including product life-cycle management software that contains computer-aided design, engineering, and manufacturing modules, which is particularly helpful for new product development projects. Computer-aided software engineering and model making also are obviously helpful in appropriate situations. Design of experiments and data-processing packages can play a role. It is obvious that office software (word processing, spreadsheet, and presentation modules) can save much otherwise tedious labor. The key point is that no one software package will solve all the problems you and your team encounter on a project, but you should adopt and make effective use of software that can help.

TYPICAL PROBLEMS

The correct way to select software is to first specify the problem you are trying to solve with it, second to locate the software that solves the problem, and third to find hardware that runs the selected software. In the case of many people who procure project management software, this process is reversed. They have existing hardware, they find some project management software that runs on it, and then they wonder why this does not solve their problem. Your challenge—and your opportunity—is to find and use project management software with which *you* can better manage *your* projects.

Finally, to reiterate a key point: If you are spending time at a computer terminal, who is managing the project team?

HIGHLIGHTS

- Computer project management software is readily available to assist a project manager in many important parts of the job.
- The use of this software will require time to train personnel.
- This software will not help the project manager to solve many of the interpersonal problems that occur.

27

Where Do You Go from Here?

We close this book with a summary of some of the key points about project functions. We describe some sources for continuing personal development, and we provide some remarks on the future of project management.

In Chapter 1, we stated that project management is a *discipline,* a word that has its semantic roots in the ideas of teaching and learning. As an individual and organizational competency, project management discipline involves the following:

1. Focusing on the important factors that drive project success
2. Avoiding the trap of staying in your personal comfort zone
3. Prioritizing the customer's requirements and conditions of satisfaction
4. Investing sufficient up-front time
5. Having the personal backbone to withstand the criticism of undisciplined, impatient people

SUMMARY

Review the following summary of the five managerial functions for your ongoing project or your next project. If you need to delve deeper into the material, turn to the chapter highlights and exhibits.

1. Define
   - Projects are not the same as tasks, programs, processes.

- Capture and document your requirements so that you have a clear, measurable understanding of the outcomes.
- Project charter.
- Have specific, measurable, attainable goals.
- Evaluate and balance competing demands by applying the Triple Constraint.

2. Plan
   - Planning is a process of learning and discovery.
   - There are many important elements of the baseline.
     - Work breakdown structure
     - Time-based critical path schedule
     - Task budgets
   - Identify risks and develop responses.
   - Develop a model of the project before automating it.

3. Lead
   - Understand your own behavior.
   - Clarify roles and responsibilities.
   - Seek other people's strengths.
   - Anticipate interpersonal conflict.
   - Negotiate work packages.
   - Communicate appropriately.

4. Control
   - Baseline, measure, assess, corrective action:
     - Project reviews help you to understand the status.
     - Get and understand reports.
     - Status first and forecast second.
     - Ask questions.
     - Apply your risk-response and issues management process.
   - Manage the change process.

5. Complete
   - Contract closure is to requirements.
   - Employ incremental documentation.
   - Carry out postcompletion work.
   - Release people back to home function and thank entire team.

As we have stressed throughout this book, develop a holistic, integrated, interrelated understanding of project management. Figure 27-1 provides a synthesized view of how the functions interact.

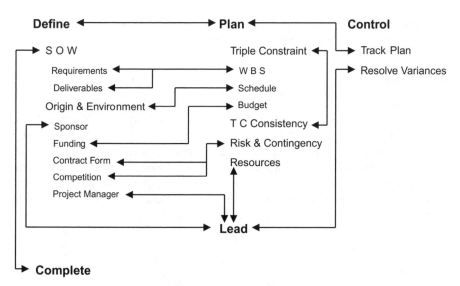

FIGURE 27-1. *Some key interactions.*

CONTINUING PROJECT MANAGEMENT SKILL DEVELOPMENT

Project management has developed considerably since the publication of the first edition of this book. Project management is now in a "golden age" with a vast and rapidly expanding knowledge base.

In evaluating individual competencies and development needs, consider three elements of performance: knowledge, skills, and desire. First, do people—whether they be project managers, project team members, functional managers, customers, or sponsors—have the knowledge? Our experience is that most of these people are still lacking knowledge of the basics, and we hope that this book fills in some of the deficiencies. Second, do people have the skill to apply the knowledge? This book describes a number of essential personal and team skills such as work breakdown structure (WBS) construction, documenting requirements, team building, communicating effectively, and so forth. Skill comes from practice, and the watchful eye of a coach or mentor is helpful. Third, do people have the desire to acquire the knowledge and develop the skill? Motivation is a complicated subject because it has both internal elements (some people are naturally more focused and energetic) and external elements (in organizations, people behave according to what they perceive as the relevant measures and rewards). There is good news: We know of many people and organizations that have gone from poor performers to high performers.

In the paragraphs that follow, we want to overview some sources of additional knowledge on the topic of project management. What works for one person may not be useful or appropriate for another. Thus, you should continue to experiment, read, and seek out other sources for professional growth.

Associations

As a "gateway" into the field, we recommend starting with the Project Management Institute (PMI), which is a nonprofit professional association with a worldwide membership. Starting from PMI's Web site, you will quickly be able to identify opportunities for certifying individuals as a project management professional (PMP), practitioner magazines, academically focused journals, career centers, seminars, congresses, and so on. PMI also has local chapters and specific interest groups so that people can network with peers and can help you to connect to professional societies focused on project management.

> Project Management Institute
> Four Campus Boulevard
> Newton Square, PA 19073-3299
> Tel: (610) 356-4600
> Fax: (610) 356-4647
> *http://www.pmi.org*

As discussed in Chapter 25, new product projects for commercial and consumer markets today require increasingly shorter development cycles. We recommend the following professional association for more knowledge about the principles, practices, resources, and knowledge of product development and management:

> Product Development & Management Association
> 17000 Commerce Parkway (Suite C)
> Mount Laurel, NJ 08054
> Tel: (856) 439-9052 or (800) 232-5241
> Fax: (856) 439-0525
> *http://www.pdma.org*

Reading

There are numerous professional books on project management, specific aspects of project management, specific software packages for project management, leadership, planning, and other related topics. In addition, there are numerous newsletters, magazines, Web blogs, and the like describing different elements of the subject.

The World Wide Web

The World Wide Web is a rapidly changing community of resources and users. One simple starting point is to use a search engine to identify sources. Two Web sites that seem to have some staying power are *Gantthead.com* and *Projectconnection.com.*

Continuing Professional Education and Academic Education

Many universities and commercial organizations provide project management seminars and workshops of varying duration. There are a growing number of university degree programs, particularly at the bachelors' and masters' levels, allowing individuals to get an excellent professional insight to the fundamentals of project management. They provide these by distance learning, on your site or on their campus, and in combinations. These vary in quality, teaching method, and subject matter, so you should identify both the instructor and the content with respect to your requirements. It is always a good idea to talk with prior participants.

Conferences and Congresses

Worldwide, several dozen conferences and congresses are offered each year, where you can meet and greet individuals interested in project management. These events can be industry-specific (e.g., automotive or drug development), application-area-specific (e.g., project risk management, project quality management, information technology, or new product development), or regional. The larger software vendors also have user-group meetings. Many large corporations have their own company events to gather employees from all around the world to participate, listen, and learn.

Coaching

Sometimes you still may find that you are "in over your head." In such a case, you may wish to identify a "consultant" who can help you to diagnose your deficiencies and develop an improvement program.

THE FUTURE OF PROJECT MANAGEMENT

Futurists make careers out of extrapolating from past events or "connecting the dots" in new creative ways. An extensive discussion of the future of project management would require a book or more on its own. Instead, we are going to mention five trends and their implications.

- *Knowledge-based work.* The saying "knowledge is power" is a cliché but nevertheless true. Much of the growth in project management in the last two decades has been among knowledge workers. While artifacts such as published project plans, specifications, and reports are important components of effectiveness, increasingly, project teams must work with intangibles. Projects are now and increasingly will be regarded as vehicles for organizational learning.

- *Continued globalization.* With growing economies in Asia and elsewhere, we are continuing to find project managers to be global travelers. In Chapter 13, we discussed the virtual organization. We expect to see a continuing increase in this practice, supported by Web-enabled tools and information systems.
- *Multitasking.* Most organizations have aggressive business goals and continue to put pressure on individuals to do "more with less." The good news is that there are many information systems to support this, but the bad news is that people still have a limited "bandwidth" for information processing.
- *Social and technological networking.* More and more individuals are part of "communities of practice" inside and outside their employers. This gives them opportunities to share best practices, recruit and be recruited, and grow personally and professionally.
- *Growing sophistication of the field.* Here is a short list of things that support the assertion the project management is becoming more mature and sophisticated: Executives and other senior managers are better understanding their responsibilities, organizations now measure their project management maturity, career paths are established, metrics are established, and there is a better understanding of how to create an effective project management culture.

A FINAL THOUGHT

In Chapter 1, we wrote that project success doesn't just "happen"; it comes from people using commonsense tools that are suited to the special nature of projects and applied in an organizational environment that accepts discipline and rigor. Few people ever read a project management book from cover to cover, and if you have made it this far, you are likely have the perseverance needed to practice successful project management. Good project management—whether individual or organizational—requires sustained energy. It is not easy but it is worth it.

It seems appropriate to close this book by reflecting on the adage, "Chance favors the prepared mind," and wish you the best of luck.

Appendix 1

Abbreviations Used in Project Management

ACO—administrative contracting officer
ACWP—actual cost of work performed
ADM—arrow diagramming method
AIN—activity-in-node
AOA—activity-on-arrow
AON—activity-on-node
B&P—bid and proposal
BAC—budget at completion
BCWP—budgeted cost of work performed
BCWS—budgeted cost of work scheduled
CCB—change (or configuration) control board
CCDR—contractor cost data reporting
CCN—contract change notice
CDR—critical design review
CDRL—contract data requirements list
CFE—customer furnished equipment
CMO—contract management office
CPFF—cost plus fixed fee
CPIF—cost plus incentive fee
CPM—critical path method
CBS—cost breakdown structure
EAC—estimate at completion
ECN—engineering change notice
ECP—engineering change proposal
EF—early finish date
EIN—event-in-node

ES—early start date
ES&H—environmental, safety, and health
ETC—estimate to complete
EV—earned value
FARS—federal acquisition regulations
FF—finish-to-finish
FFE—fuzzy front end
FFP—firm fixed price
FP—fixed price
FS—finish-to-start
G & A—general and administrative
GIGO—garbage in, garbage out
IR & D—internal research and development
LAN—local area network
LF—late finish date
LS—late start date
MBC—management by commitment
MBO—management by objectives
MBR—management by results
MBWA—management by walking around
NPD—new product development
ODC—other direct cost
OBS—organizational breakdown structure
PBS—product breakdown structure
PDM—precedence diagramming method
PDR—preliminary design review
PERT—program evaluation and review technique
PM—project (or program) manager (or management)
PMBOK—Project Management Body Of Knowledge
PMI—Project Management Institute
PMO—project management office, program management office
PMP—Project Management Professional
PO—purchase order
PWA—project work authorization
QA—quality assurance
QC—quality control
QFD—quality function deployment
R&D—research and development
R,D&E—research, development and engineering
RFP—request for proposal
RFQ—request for quotation
S&G—stages and gates
SOW—statement of work

SS—start to start
T&M—time and material
WBS—work breakdown structure
WO—work order
3D—three-dimensional

Appendix 2

Glossary of Project Management Terms

Activity. A single task within a project.

Actual cost of work performed (ACWP). A term in the earned-value management system for the costs actually incurred and recorded in accomplishing the work performed within a given time period.

Arrow diagramming method. A type of network diagram in which the activities are labeled on the arrows.

Bar chart. A scheduling tool (also called a *Gantt chart*) in which the time span of each activity is shown as a horizontal line, the ends of which correspond to the start and finish of the activity, as indicated by a date line at the bottom of the chart.

Baseline. The approved project plan, used to establish monitoring and control activities. Often used with a modifier, such as *scope baseline, cost baseline,* or *schedule baseline.*

Bid/no-bid decision. The decision whether or not to submit a proposal in response to a request for proposal.

Bottom-up cost estimating. The approach to making a cost estimate or plan in which detailed estimates are made for each task shown in the work breakdown structure and summed to provide a total cost estimate or plan for the project.

Budgeted cost of work performed (BCWP). A term in the earned-value management system for the sum of the budgets for completed work packages and completed portions of open work packages, plus the appropriate portion of the budgets for level of effort and apportioned effort.

Budgeted cost of work scheduled (BCWS). A term in the earned-value management system for the sum of budgets for all work packages, planning packages, and similar items scheduled to be accomplished (including in-process work packages), plus the amount of level of effort and apportioned effort scheduled to be accomplished within a given time period.

Buy-in. A process of gaining stakeholder commitment and support for the project. Alternatively used to indicate the practice of pricing a bid very low to ensure that the offeror will win the project.

Chart room. A room filled with planning and status documents displayed as charts, typically hung on the walls of the room; used on large projects.

Commitment. An obligation to pay money at some future time, such as a purchase order or travel authorization, which represents a charge to a project budget even though not yet actually paid. Alternatively used to mean that stakeholders support the project with resources, time, and energy.

Contingency. See *reserve.*

Cost work breakdown structure. A decomposition of the budgeted cost of the project into smaller elements.

Cost plus fixed-fee (CPFF) contract. A form of contractual arrangement in which the customer agrees to reimburse the contractor's actual costs, regardless of amount, and in addition pay a negotiated fee independent of the amount of the actual costs.

Cost plus incentive-fee (CPIF) contract. A form of contractual arrangement similar to CPFF except that the fee is not preset or fixed but rather depends on some specified result, such as timely delivery.

Critical path. In a network diagram, the longest path from start to finish or the path without any slack; thus, the path corresponding to the shortest time in which the project can be completed.

Customer-furnished equipment (CFE). Equipment provided to the contractor doing the project by the customer for the project and typically specified in the contract.

Documentation. Any kind of written report, including such items as final reports, spare parts lists, instruction manuals, test plans, and similar project information.

Dummy activity. An activity in a network diagram that requires no work, signifying a precedence condition only.

Early finish. In a network diagram schedule, the earliest time at which an activity can be completed.

Early start. In a network diagram schedule, the earliest time at which an activity can be started.

Earned-value management. A performance management approach that integrates work scope, budgeted cost, and schedule.

Event. An occurrence. In scheduling, an event is an activity that consumes no resources and has no duration and typically is portrayed as a diamond shape on graphics.

Firm-fixed-price (FFP) contract. A contractual form in which the price and fee are predetermined and not dependent on cost.

Fixed-price (FP) contract. Same as firm fixed price.

Float. In schedule network analysis, the amount of time on any path other than the critical path that is the difference between the time to a common node on the critical path and the other path.

Functional organization. The form of organization in which all people with a particular kind of skill (such as engineering) are grouped in a common department, reporting to a single manager for that particular functional specialty.

Late finish. In a network diagram schedule, the latest time at which an activity can be finished.

Late start. In a network diagram schedule, the latest time at which an activity can be started.

Matrix organization. The form of organization in which there is a project management functional specialty as well as other functional specialties and where the project management function has responsibility for accomplishing the project work by drawing on the other functional specialties as required.

Milestone. A major event in a project, typically one requiring the customer to approve further work.

Network diagram. A scheduling tool in which activities or events are displayed as arrows and nodes in which the logical precedence conditions between the activities or events are shown.

Periodic review. Any kind of project review conducted on a periodic basis, most commonly a monthly project review.

"The Plan." A document or group of documents that constitutes all the plans for the project, frequently contained in a notebook or series of notebooks.

Planning matrix. A matrix in which planned activities are listed on one side (usually the left) and involved people or groups are listed across a perpendicular side (usually the top) and where involvement of a particular individual or group in a particular activity is signified by a tic mark where the row and column intersect.

Precedence diagramming method. A type of network diagram in which the events and activities are labeled in the nodes or boxes and are linked by logical relationships that show the ordering of tasks.

Problem scope. See *requirement*.

Product scope. The features and functions of the result, determined by some kind of design activity stimulated by analysis of the customer's requirements specifications.

Program. A group of projects managed in a coordinated way to obtain benefits and control not available from managing them individually.

Progress payments. Payments made to the contractor by the customer during the course of the project rather than at the end of the project, the terms of which are specified in the contract.

Project. A kind of work that is temporary, unique, and progressively elaborated. The project is the work that produces the result.

Project organization. The form of organization in which all or nearly all the people working on a project report to the project manager.

Project plan. The outputs of the project planning process, typically intended as a resource for the project team and for use on controlling activities.

Project scope. See *work scope.*

Proposal. A document that an organization submits to a prospective customer that describes work the organization is offering to do.

Request for proposal (RFP). A document issued by one organization to another organization (or to several other organizations) describing work that the issuer wishes to have undertaken by the recipient(s) and inviting the recipient(s) to respond with a proposal.

Request for quotation (RFQ). Similar to a request for proposal, except that the desired items to be procured are stock or catalog items, and only price and delivery time need be proposed.

Requirement. A capacity needed by a user to solve a problem. A capability that must be possessed by a system. Requirements reflect the wants and needs of the customer but are not necessarily the same. Requirements are an agreement between the customer and the project delivery organization.

Reserve. An amount of design margin, time, or money inserted into the corresponding plan as a risk management response factor to accommodate risk events.

Risk event. A discrete occurrence that may affect the project positively or negatively.

Risk avoidance. A risk-response technique where the project team selects strategies intended to avoid exposure to a risk event.

Risk acceptance. A risk-response technique where the project team selects strategies intended to manage the risk event after it occurs. Contingency is an example of a risk acceptance strategy.

Risk mitigation. A risk-response technique where the project team selects strategies intended to reduce the probability or impact of a risk event.

Risk transfer. A risk-response technique where the project team selects strategies intended to transfer the risk to another party that is better positioned to accept and manage the risk.

Scope. A process of determining what requirements and design elements are to be considered by the project as part of its work activities.

Scope creep. Adding features or functionality without addressing the effects on time and resources or without customer agreement.

Slack time. See *float.*

Statement of work (SOW). A narrative description of requirements that often provide more detail on the expected work, services, or results.

Subcontractor. An organization, usually a company, working for another organization on some aspect of the project for which the other organization is under contract.

Support team. A term used in this book to designate the personnel working on a project who do not report to the project manager administratively.

Task. A smaller work activity within a project.

Task force. An *ad hoc* group designated to cope with a project, similar to a project organization, although frequently staffed with personnel on part-time assignment; usually adopted by a functional organization having only one project or at most a few projects at any given time.

Time compression. The act of reducing the planned time for an activity, accomplished perhaps by adding unplanned staff or using overtime.

Time and material (T&M) contract. A contractual form in which the customer agrees to pay the contractor for all time and material used on the project, including a fee as a percentage of all project costs.

Top-down cost estimating. The approach to making a cost estimate or plan in which judgment and experience are used to arrive at an overall amount; usually done by an experienced manager making a subjective comparison of the project to previous, similar projects.

Topical review. Any kind of project review devoted to a single topic, such as a final design review or a manufacturing review.

Triple Constraint. A framework for evaluating competing demands. In this book, the Triple Constraint is presented as the performance specification, the time schedule, and the monetary budget.

Venture organization. The form of organization used in some large organizations where a three- or four-person team, itself functionally organized, is established within the larger organization to develop and commercialize a new product.

Work breakdown structure (WBS). A family tree, usually product-oriented, that organizes, defines, and graphically displays the hardware, software, services, and other work tasks necessary to accomplish the project objectives.

Work scope. The work done to create the result that satisfies the requirements.

Examples of Planning Checklists for Project Managers

1. Hardware engineering or study projects

system review
system approval
system test criteria
system test plan
detailed hardware specification
customer furnished equipment
power requirement plan
weight control plan
breadboard design
breadboard fabrication
breadboard test
block diagram
schematic diagrams
circuit diagrams
conceptual design review
preliminary design review
critical design review
final design review
prototype design
prototype fabrication
prototype test

design freeze
drawings freeze
functional designs
system logic design
optical design
mechanical design
electronic design
thermal design
subsystem hardware
implementation
subsystem software
implementation
subsystem integration
subsystem review
subsystem approval
subsystem test criteria
subsystem test plan
make/buy decisions
long lead items
special test equipment
commercial test equipment
calibration of test equipment
software tests

data reduction plan
operational software
subsystem cabling
system cabling
installation planning
experimental development plan
support plans
support instrumentation facilities
training plans
repair facilities and requirements
inspection
preshipment review
customer inspection
customer acceptance
preparation for shipment
shipment
customer support
qualification test
flight acceptance test
launch support
mission support
personnel recruitment
personnel reassignment
documentation
project plan
integrated schedule
functional requirements
    document
environmental requirements
    document
environmental test specifications
environmental test procedures
environmental test reports
interface control
safety plan
configuration control plan
failure mode and effect analysis
reliability and quality assurance
    plan
development test plan
acceptance test procedure
calibration plan
ground data-handling plan
experiment development plan
expendables consumption

engineering drawings and
    drawing list
electronic parts acquisition and
    screening plan
materials documentation
manufacturing release
periodic reports (for example,
    monthly)
special reports
final reports
instruction manuals
reviews
manufacturing review
management review
critical design review
preshipment review
internal project reviews
subcontractor progress reviews
customer reviews

2. **Programming projects**
applications requirements
systems requirements
system inputs
system outputs
detailed architectural design
design specifications
functional specifications
security plan
system test and acceptance
    specifications
feasibility studies
file and data requirements
cost/benefit analysis
system design
program design
system conversion plan
shipment and delivery
turnover to operations
postimplementation reviews
supplies
training aids
review procedures
hardware requirements
personnel capabilities

milestone reviews
parts list
milestone documents
product specifications
project plan
operator instructions
user instructions
library
project index
change control system
data base administration
technical interface manuals
hardware reference manuals
release information
systems reference manual

3. Construction projects

project work order plan
personnel assignment
client review meetings
public involvement meetings
environmental report meetings
    with regulatory agencies
population projections
residential load projections
commercial load projections
industrial load projections
field checking
stream gauging records
stream gauging measurements
meteorological records
meteorological measurements
environmental baseline data
environmental sampling
    programs

ground surveys
photogrammetric surveys
geotechnical explorations
permits
access roads
hydrology
water quality and pollution
    studies
stability analysis
transient analysis
flood control studies
mathematical modeling
easement and permit drawings
    and descriptions
wiring diagrams
piping diagrams
construction cost estimates
estimated operation and
    maintenance costs
life cycle cost analysis
economic studies
value engineering
rate study
notice and instructions to
    bidders
bid schedules
contract documents
recommendations of awards
shop drawing review
construction surveys
construction engineering
construction observation
construction record drawings

Index

Costs →

- Breakout prices into materials, equipment, personnel with budgeted items accordingly
- look at the # of items priced
- establish associated costs
- Look at alternatives
- choose the min cost alternative